FACILITY INTEGRITY MANAGEMENT

FACILITY INTEGRITY MANAGEMENT

Effective Principles and
Practices for the Oil, Gas,
and Petrochemical Industries

MICHAEL GUY DEIGHTON

Amsterdam • Boston • Heidelberg • London
New York • Oxford • Paris • San Diego
San Francisco • Singapore • Sydney • Tokyo
Gulf Professional Publishing is an imprint of Elsevier

ELSEVIER

G | P
‾‾‾
P | ᐁ

Gulf Professional Publishing is an imprint of Elsevier
50 Hampshire Street, 5th Floor, Cambridge, MA 02139, USA
The Boulevard, Langford Lane, Kidlington, Oxford, OX5 1GB, UK

Notices

Knowledge and best practice in this field are constantly changing. As new research and experience broaden our understanding, changes in research methods, professional practices, or medical treatment may become necessary.

Practitioners and researchers must always rely on their own experience and knowledge in evaluating and using any information, methods, compounds, or experiments described herein. In using such information or methods they should be mindful of their own safety and the safety of others, including parties for whom they have a professional responsibility.

To the fullest extent of the law, neither the Publisher nor the authors, contributors, or editors, assume any liability for any injury and/or damage to persons or property as a matter of products liability, negligence or otherwise, or from any use or operation of any methods, products, instructions, or ideas contained in the material herein.

Library of Congress Cataloging-in-Publication Data
A catalog record for this book is available from the Library of Congress

British Library Cataloguing-in-Publication Data
A catalogue record for this book is available from the British Library

ISBN: 978-0-12-801764-7

For information on all Gulf Professional Publishing
visit our website at http://store.elsevier.com/

DEDICATION

This book is dedicated to my gorgeous wife Jenni, who has supported me throughout my career, and my amazing children, Isaac, Isabelle and Guy.

CONTENTS

ACKNOWLEDGMENTS

I would like to take a moment to express my gratitude to the following individuals who have supported me and helped to make this project possible.

A special thank you to the team at Elsevier, in particular Katie Hammon for believing in me and this project from the outset, Kattie Washington for her support during the 12 months of writing and development, and Anusha Sambamoorthy for helping me to fine tune and complete the manuscript.

I was fortunate enough to have had exceptional teachers and role models throughout my career who have guided me and inspired me, which has helped to prepare me for this project thank you to: James Moore, Dr. Roger Brooks, Mike Reading from ConocoPhillips, John Shanahan and Dave Wager from Dupont, K.S Sebastian from ZADCO, Dr. Ian Potts and Dr. Jack Hale from the University of Newcastle, and J. Borwell from Stokesley Collage.

I owe a debt of gratitude to my mother, Carole, and my father, Alan. They have made countless sacrifices to ensure I received a well-rounded education and supported me all the way in my career. I have been taught to always strive for excellence in everything I do ("no job is too big…"). Thank you, Mum and Dad.

Writing a book has taken its biggest toll on my family. This project has meant less time for us to spend together over the many weekends and late nights working. My immense gratitude goes to my amazing wife Jenni (WW) and our beautiful children Isaac, Isabelle and Guy.

Michael Guy Deighton, 2015

PREFACE – ABOUT THE AUTHOR

This book focuses on the practical applications of a collection of facility integrity management, operations and reliability principles which have been tried, and tested on oil and gas and petrochemical facilities.

The author Michael Guy Deighton has over 20 years of diverse experience in the oil, gas and petrochemical sectors, working for the world's leading companies within the fields of facility integrity management, maintenance, operations and capital projects and has led some of the pioneering facility integrity development and implementation projects for world-class international oil and gas companies. He has held the titles of Senior Integrity Consultant, Reliability Champion, Maintenance Manager, Site Engineering Manager, Operations Manager, and Project Director. He is a Fellow of the Institute of Mechanical Engineers, with master's degrees in mechanical engineering and business management.

The author has had the opportunity to work as consultant in developing new integrity management systems for leading organisations. As a result of this knowledge and experience, he has developed an all-encompassing integrity excellence model that provides an in-depth account of integrity, reliability, operations and maintenance processes for petrochemical and oil and gas facilities personnel. The model is founded on the premise that there are synergies behind the natural interdependencies of the elements that make up facility integrity management and operations. With one element missing, the whole process is vulnerable and at risk of failure. The author's integrity excellence model approaches facility integrity from a holistic perspective and integrates these key concepts, structuring them into a seamless and easy-to-use guide for graduates, engineers and facility managers.

DEFINITIONS AND NOMENCLATURE

Acceptable risk A level of risk below which risks are so low that they may be safely eliminated from further consideration. No need to demonstrate ALARP.

ALARP The risks are reduced to a level that is as low as reasonably practicable.

ASME American Society of Mechanical Engineers

Audit A systematic and independent examination of activities and systems to determine whether they are implemented effectively to achieve the set objectives.

Availability The proportion of the total time that a component, equipment, or system is performing in the desired manner.

Checklist A comparative method for hazard identification based on lists of failure modes compiled from experience.

CMMS Computerized Maintenance Management System

Consequence The outcome of a hazardous event.

DMAIC Six Sigma project execution methodology: "Define, Measure, Analyze, Improve, Control"

DMS Document Management System

Downtime A period of time that a machine or system is offline or not functioning, usually as a result of either system failure or routine maintenance. The opposite is uptime.

EMOC Equipment Maintenance and Operating Card

EMOP Equipment Maintenance and Operating Plan

Event Tree A method for representing graphically, using inductive reasoning, the intermediate and final outcomes arising from a given initial event.

Explosion A sudden release of energy which causes a pressure discontinuity or blast wave.

FEF Facility Equipment Failure

Fault Tree A method for determining in a systematic manner, using logical relations and deductive reasoning, the primary causes of a specified event.

FIEM Facilities Integrity Excellence Model

FI&R Facility Integrity & Reliability

Frequency The number of occurrences per unit time.

Functionality The ability of a system to perform its specified role. This may be characterized and demonstrated by identifying critical functional parameters.

GIS Global Information Systems

Hazard	A set of conditions or actions with a potential for human injury, damage to property, and damage to the environment or a combination of these.
Hazard Analysis	The determination of consequences associated with a given hazard.
HAZOP	Hazard and Operability Study; A method for identifying hazards using guide words to establish deviations from the design intent.
HVAC	Heating, Ventilation and Air Conditioning.
ICMS	Inspection and Corrosion Management System
Integrity Limit	Operating envelope limits for equipment and systems.
ISO	International Organization for Standardization
IT	Information Technology
KPI	Key Performance Indicators
LPG	Liquefied Petroleum Gas
LPO	Lost Profit Opportunity
MCC	Motor Control Center
Mitigation	Means taken to minimize the consequences of a major accident to personnel and the installation after the accident has occurred.
MMS	Maintenance Management System
MTBF	Mean time between failures is the mean (average) time between failures of a system.
MTBR	Mean time between repair is the mean (average) time between repair of a system.
MTTR	Mean time to repair is the mean time it takes for the equipment or system to repair and reinstate the equipment back into service after a failure has occurred.
NDT	Non-Destructive Testing
OEM	Original Equipment Manufacturer
P&ID	Piping and Instrumentation Diagram
Performance Standard	A statement, expressed in qualitative or quantitative terms, of the performance required of a system, item of equipment, person or procedure, and which is used as the basis for managing the hazard.
PHA	Process and Hazard Analysis
PMI	Positive Material Identification
Prevention	Means (engineering as well as management) intended to prevent the initiation of a sequence of events which could lead to a hazardous outcome.
Probability	The chance of occurrence of a specified event (expressed on a scale from 0 to 1).
Redundancy	The performance of the same function by a number of identical but independent means.

Reliability	Reliability is concerned with avoiding failures of equipment and processes by proper design and careful operation of the equipment by trained personnel in a specified environment for a given time interval.
	Is the probability that a component or system will perform a required specified function.
Residual Risk	The level of risk that remains when all reasonable steps to prevent, control or mitigate risk have been taken.
Risk	The product of the likelihood or probability of an event being realized and its consequences.
Risk Analysis	The quantified calculation of probabilities and risks without any judgements about their relevance.
Risk Assessment	A systematic analysis of risk in terms of its probability and consequences and its significance in an appropriate context.
Risk Management	The process whereby decisions are made to accept a known or assessed risk and/or the implementation of actions to reduce the consequences or probability of occurrence.
RTF	Maintenance strategy involving deliberately adopting a "Run To Failure" approach for noncritical Plant Equipment.
Safety Management System	The application of organizational procedures and methods to meet the specified safety objectives.
SAP	Material requirements planning and ordering software. German software company abbreviation: "Systeme, Anwendungen, Produkte in der Datenverarbeitung" translates to "Systems, Applications and Products" in the area of data processing.
TMT	Transition Management Team
Uptime	A measure of the time a system has been "up" and running. It came into use to describe the opposite of downtime (times when a system was not operational).
Variance	A deviation from the equipment normal operating performance.

FIGURE LISTING

TABLE LISTING

CHAPTER 1

Introduction

Contents

1.1 INTRODUCTION – FACILITY INTEGRITY MANAGEMENT

The oil, gas and petrochemical sectors spend billions of dollars every year to maintain facility integrity. If done well, facility integrity management can have a positive impact on a company's profit and loss account, prolong the life of the facility and optimize the operating efficiency and effectiveness of processes and equipment. Conversely, if done poorly, it can reduce large blue chip companies to bankruptcy.

Facility integrity management is the means of ensuring that the people, systems, processes and resources that deliver integrity are in place, in use and fit for purpose over the whole life cycle of the facility. It is a process that requires an optimized balance between resources and output; ensuring facility integrity management is well structured and managed is key to its success in terms of a well-run and managed facility and a good return on facility production output.

Facility Integrity Management
http://dx.doi.org/10.1016/B978-0-12-801764-7.00001-2

This book is about how the concepts within facility integrity management can improve the performance of a facility, making substantial improvements to the safety, operational effectiveness, and ultimately profit of an oil, gas and petrochemical production facility. It provides a detailed explanation of the key principles and processes with easily referenced material that has been tried and tested successfully in the field.

1.2 CHANGING INDUSTRIAL LANDSCAPE

Producing high-quality products at competitive prices is paramount to maintaining a successful business in today's ever-evolving industrial landscape. Oil, gas and petrochemical operators are constantly challenged to ensure their facilities are safe and environmentally friendly and their production equipment is running smoothly and is available whenever required. In order to satisfy customers' needs, operators are constantly looking for new, improved ways to ensure equipment and systems are available and operating as lean and efficiently as possible, and at the same time reducing cost.

Over the past few decades the approach to integrity management of petrochemical and oil and gas facilities has changed dramatically, perhaps more so than most other industrial markets. This change has come about through several key drivers in the industry. Probably the most fundamental driver has been the need to make a step change in the approach to process safety in response to a string of process facility disasters. This is followed by the fact that there are now many more operators in the market than in the past and in order to maintain profit margins, operators are required to be much leaner, lowering their overhead costs. Operators have also found that there has been an increasing demand for higher specification products, mainly to cope with the more stringent legislative requirements but also to meet the end user demands, such as technological advances in automotive and aviation products.

These changes have had a profound impact on oil, gas and petrochemical facility operators. This has led to an increase in the demand on processes and equipment to perform more complex processes longer and more efficiently, and also to come up with equipment and new product development, leading to new processes and equipment.

In the face of this torrent of change, facility managers are looking for new, more cost effective, ways to improve equipment reliability, maximize uptime, and optimize maintenance and operations efforts in order to reduce unscheduled downtime in their approach to operating and maintaining their facilities.

Many facility operators have responded by going back to first principles and rethinking the way their integrity and maintenance groups should operate. Facility integrity and reliability (FI&R), maintenance, and operations departments in the past have traditionally operated in silos – that is, operated as independent units, effectively disconnected from each other to a large extent. This has inadvertently put up barriers around the core integrity departments and ultimately disrupted the flow of information between the various interfaces within the facility, from the facility leadership team to the various stakeholders, having a negative impact on facility operations.

As a result of the numerous major oil, gas and petrochemical facility incidents that have caught the headlines over the years, operating companies are considering integrity and reliability as a standalone department that works alongside maintenance and operations departments with the primary goal of ensuring safe operation and optimum availability of facility processes and equipment. In order to illustrate this point, I draw upon a case study which occurred at the Humber Refinery in the UK on 16th April 2001, where there was a serious incident resulting in an explosion and fire at the Refinery Saturated Gas Plant; the incident was investigated by the UK Health and Safety Executive [1.1]. One of the key outcomes of the report was a reorganization of the Refinery Organization Structure with emphasis on the integrity department. This case study is explored in more detail, in order to draw out the key learning points, in Chapter 2.

A robust and well-implemented facility integrity management system is essential to ensure coordination between these departments in order to ensure all aspects relating to integrity of the facility have been considered and adequately addressed. The development and implementation of an effective integrity management system requires a concerted and dedicated effort across the facility organization.

1.3 THE ROLE OF FACILITY INTEGRITY

The fundamental role of facility integrity in the operation of oil, gas and petrochemical facilities is to:
1. Maximize the availability of the facility.
2. Provide integrity assurance for the facility.

The role of maintenance and integrity has never been more critical to the oil, gas and petrochemical industries. In addition to providing confidence and assurance to facility owners that operational risks can be identified and managed, the facility can now drive towards achieving operational excellence by optimizing the performance and extending the life of equipment and systems.

Maintenance and integrity have a key role to play across all stages of the project life cycle, not just during the operating stage but also during the design stage where many of the critical process safeguards and availability of equipment and processes are established. We shall now consider each of the fundamental roles of maintenance and integrity in turn.

1.3.1 Ensure the Availability of all Facilities

Availability is a measure of facility *uptime* or, in other words, how often facility equipment and systems are alive and function as they were designed to. Ensuring the availability of the facility in today's market is critical in order for business to be successful. Today's global marketplace is fueled by an ever-increasing demand for oil and gas commodities to feed the growing global energy demand.

Since 1952, the *BP Statistical Review of World Energy* has provided objective global data on energy markets. This longevity and robust reporting helps put today's picture into context to help us understand how the world around us is changing. The *BP Statistical Review of World Energy* for June 2014 presents the global demand for crude oil and natural gas, among other energy sectors.

The illustration in Figure 1.1 shows the production and consumption by region over the last 25 years. In 1988 the world demand was approximately 65 million barrels per day, growing to over 90 million barrels per day in 2013. In 25 years global consumption has grown by nearly one and a half times.

Similarly, the trend for natural gas was approximately 1800 billion cubic meters per day in 1988 and has grown to nearly 3500 billion cubic meters per day in 2014, almost doubling in growth in 25 years.

Coupled with the natural growth rates of oil and gas commodities and derivatives, there has also been a demand on oil, gas and petrochemical operators to improve product specifications in order to meet the ever more stringent compliance requirements in line with global environmental standards.

A key driver of this change is through global emissions reduction initiatives. Emissions standards focus on regulating pollutants released by automobiles and emissions from industry, power plants and small equipment such as diesel generators. There are numerous emission standards that have all followed this theme, which has primarily grown out of the Kyoto protocol

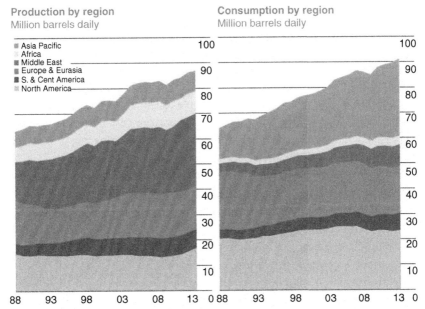

Figure 1.1 *BP statistical review of world energy 2014 – production and consumption by region.*

[1.2], including the US Environmental Protection Agency (EPA), Clean Air Act 1970, UK Health and Safety Executive, among many others.

In addition to product development through compliance-driven regulations, a number of new products have come into the mix in recent years. Most notable are alternatives to lead-based fuels for the automotive industry and synthetic fluids and advanced lubricant oils, which have been driven through technology-pull developments in the automotive industry.

The net result of all this change is that there is an ever-increasing demand for oil, gas and petrochemical commodities, constant requirements for improving product specifications, and new products. As a result oil, gas and petrochemical operators have been forced to embark on a journey to make step change approaches in their quest to remain competitive and compliant and develop new ways to increase productivity, maximize process equipment uptime, ensure world class safety standards and altogether reduce operations and maintenance cost.

The secondary effect is an increase in the number of and complexity of manufacturing processes that need to operate with a high degree of availability and maintenance.

Oil, gas and petrochemical operators have responded by taking a critical look at the way they approach integrity management, reliability and maintenance with the intention of radical change.

1.3.2 Safeguard the Integrity of all Facilities

Over the last 50 years or so there have been a number of major petrochemical, oil and gas disasters resulting in loss of life, severe environmental impacts and property damage, providing reasons for the public to be wary about oil, gas and petrochemical facilities.

There are many underlying root causes leading to an initiating event; however, in many cases, the approach to management of integrity has been identified as a major cause. A number of recent disasters bear strong evidence to this proclamation. A brief overview of two of the major oil, gas and petrochemical disasters that resulted in step change approaches to integrity management is given in the following sections.

1.3.2.1 The Piper Alpha Disaster

The world's deadliest-ever offshore oil rig accident happened on July 6, 1988, killing 167 out of 228 men in 22 minutes.

A report following the disaster by Scottish Judge Lord Cullen concluded that the operator, Occidental Petroleum, had used inadequate maintenance and safety procedures, resulting in the largest manmade catastrophe. A series of grave mistakes was made in design and operation that ultimately led up to the incident. A key contributor was the modifications completed to the initial design, including a new gas compression process which was located near the existing central control room, perceived to be one of the most dangerous production areas on the platform. There clearly was little effort to consider and safeguard against the associated risk, in terms of management of change.

During operation, the operator also decided to keep the platform operational and producing oil and gas as it made a series of major construction, maintenance and upgrade changes, which increased the risk factor.

There was a distinct lack of communication during shift changeover, which meant that the rig workers were unaware that they should not use certain piping, which was under maintenance at the time. This piping had been temporarily sealed with no safety valve. Ultimately, gas leaked out at high pressure, ignited and exploded. The protective firewalls failed because they were designed to withstand fire and not explosions, as fundamentally the original design intent of the platform was for oil production and not gas production, which is potentially far more hazardous.

Lord Cullen's report included 106 recommendations to the industry, all of which were accepted. The outcome of this disaster has had a profound effect on the oil and gas industry and sparked a complete overhaul of the governing safety regulations, which included Offshore Installations (Safety Case) Regulations 1992. This meant that all operators were now required to present a safety case to the Health and Safety Executive (HSE) agency for oil and gas production facilities [1.3].

1.3.2.2 The Flixborough Disaster

On June 1, 1974 a large explosion at the Nypro Chemical facility at Flixborough in the UK resulted in 28 fatalities and severe property damage to the facility and surrounding area. The devastation was vast; fires started on-site after the explosion were still burning some 10 days later with around 1,000 buildings within a mile radius of the site in Flixborough and neighboring villages damaged.

The Nypro Facility produced caprolactam from cyclohexanone, a highly flammable material, which was ultimately used in the manufacture of nylon.

Prior to the explosion, it was uncovered that a vertical crack in one of the reactors on the facility was leaking cyclohexane. The facility was subsequently shut down to investigate, which resulted in a serious problem identified with the reactor. A decision was therefore made to remove the problematic reactor from service and install bypass connecting pipework to connect the adjacent reactors in order to continue production.

During the late afternoon of June 1, 1974, the bypass system ruptured, which resulted in the escape of a large quantity of cyclohexane. The cyclohexane formed a flammable mixture and subsequently found a source of ignition; shortly after there was a massive vapor cloud explosion.

The formal report issued by the court of inquiry identified a number of failings. The report noted that the facility was "well designed and constructed" but the bypass system modification was implemented without a full assessment of the potential consequences. Only limited calculations were undertaken on the integrity of the bypass line. No calculations were undertaken for the dog-legged shaped line and no drawing of the proposed modification was produced. The bypass line had also not undergone thorough testing and analysis including a pressure test.

The report noted that there was a distinct lack in the capacity and competence of personnel on the facility, especially during the modification of the bypass line. It was noted that at the time of the disaster no professionally qualified engineers were in the works engineering department.

The control room was occupied by 18 staff, all of whom lost their lives as a result of the windows shattering and the collapse of the roof. The design of the layout, including the control room, was not done based on consideration of a major disaster happening instantaneously. Furthermore, the structural design of the control room did not allow for major hazard events.

The report highlighted a number of observations and lessons to be learned. Proper management of change needs to be employed during facility modifications, which should be designed, constructed, tested and maintained to the same standards as the original plant.

There was no hazard assessment or Hazard and Operability study (HAZOP) conducted during the modification to understand the implications of the change. The incident happened during start up, a particularly stressful environment where critical decisions were made. For example, there was a shortage of nitrogen for inerting. This would inhibit the venting of off-gas as a method of pressure control and/or reduction.

The disaster was met with public outcry and was criticized for the lack of change management and has been instrumental in a more detailed and systematic approach to process safety in the UK. The emphasis was on fast restart of the plant after the initial leak and not on the investigation of the new modification and of any implications of this modification on future production and safety [1.4].

1.3.2.3 The Effect on the Oil, Gas, and Petrochemical Industries

Many of the major oil, gas, and petrochemical incidents arose from the cumulative effect of a range of errors and vulnerabilities introduced throughout the life cycle of the facility through design, construction, operation and maintenance. In many cases, there is a considerable amount of fragmentation between the different facility departments, including operations and maintenance; in many circumstances these departments are outsourced, leading to further fragmentation.

Oil, gas and petrochemicals are hazardous substances and most of the equipment used in the facilities is there to contain and manage the hazards. Keeping oil, gas and petrochemical facilities available and in good working order is critical in order to eliminate loss of containment and ultimate disasters.

The two main causes of loss of containment are either failure of the facility equipment or human factors, and may be initiated through a number of mechanisms. Facility equipment failure (FEF) may include poor process design, incorrectly specified equipment, corrosion or erosion mechanisms, overpressure or over temperature, stress or fatigue or vibration mechanisms, defective equipment and degraded or aged equipment. Personnel failure (PF) may include noncompliance with procedures, inadequate isolation during operation or maintenance, dropped objects or impact to equipment and incorrect installations.

The catastrophic Piper Alpha platform disaster in 1988 shocked the world and woke up the oil and gas industry. The Lord Cullen report concluded that there was far too little understanding of oil and gas operational hazards and management of their associated risks. The industry came to the realization that there is a need to develop a systematic and methodological way of assuring control over these hazards.

As a result of Lord Cullen's report, the Offshore Installations Safety Case Regulations came into force in 1992. The law dictates that owners and operators of all fixed and mobile offshore installations in UK waters must provide a safety case to the HSE. The safety case must demonstrate that the company has facility integrity management systems in place. The systems must have identified risks and reduced them as much as is practically possible. The safety case must also have made provisions for safe evacuation and rescue. As of 1995, each safety case for all installations falling under these regulations had to be submitted and accepted by the HSE.

The Piper Alpha Disaster had a landmark effect on the oil and gas industry. The industry's approach to the management of facility integrity radically changed. Oil and gas owners and operators conducted immediate and comprehensive assessments of their facilities and corresponding facility integrity management systems. Offshore operators reviewed their strategies and invested over a billion USD to address shortcomings in integrity management of their facilities. This was the nucleus for the change in culture, leading more towards integrity management and safe operations.

1.4 FACILITY INTEGRITY MANAGEMENT SYSTEMS

1.4.1 Does This Look Familiar…?

Before we move forward let us take a moment to consider what some of the day-to-day scenarios may look like in the absence of a facility integrity management system:

- Aging equipment at extended life and operating at maximum levels
- Gaps in management systems and fragmented reliability programs
- Acceptance of equipment and facility processes operating with substandard performance

- No or limited attention to management of knowledge of equipment and process data
- Unplanned shutdowns are common
- Silo mentality across the organization with no or limited interdepartmental cooperation or strategic direction
- Dashboards are prepared and issued on an ad hoc basis and information is not reliable
- Changes in facilities operation and modifications to equipment are not documented nor managed through a robust system

Effective management of facility integrity can bring about step change improvements on a facility. These improvements can be seen across the breadth of a facility, even areas that may traditionally not be realized and the benefit not fully understood. These may include improvements to production through equipment and systems optimization, minimizing unplanned shutdowns and associated lost profit opportunities, optimizing maintenance and inspections, workshop efficiency improvements, right sizing of the facility workforce, all resulting in cost savings.

Facility integrity also includes compliance with regulatory and company requirements being in place. In this desired state, we will find a very different picture. Some of the following scenarios are likely to be common:

- A facility operating with zero unplanned shutdowns and incidents
- An organizational culture that take a keen interest in identifying and resolving process upsets and equipment malperformance
- A competent, experienced and dedicated workforce
- Continuous improvement and optimization of facility process work flow
- A facility dashboard with proactive performance indication
- Synergy between the numerous facility working groups and teams
- Knowledge-based organization with a reliable historian of plant performance and operation data

The development of an effective system that can manage the integrity of a facility requires a structured and dedicated approach. A facility integrity management system does not exist in isolation. Successful implementation requires a unified approach across the facility organization with a common understanding of integrity management and the criticality of each employee's role.

1.4.2 Facility Integrity "Excellence"?

We have discussed that the fundamental role of maintenance and integrity is to maximize facility availability and safeguard integrity assurance; now taking this one step further, let's consider what needs to be done to achieve facility integrity "excellence."

Striving to achieve excellence in facility integrity management would certainly result in the aim to maximize the availability of the facility in order to achieve target production. It would also aim to safeguard integrity assurance in order to eliminate integrity-related incidents. Excellence must also aim to maintain the facility in a fit-for-service condition with all equipment and systems functioning as per their original design intent.

Since it is all too common that many facilities are functioning beyond their original intended design life, facility integrity excellence should also consider the assessment of remnant life of equipment and systems with the intention of prolonging facility life.

Ensuring there is a practical consideration for cost is also a key attribute of a facility integrity management system. There are no blank checks and all operators are competing in the industry to make money. Each element within a facility integrity management system needs to consider value for money and minimize cost where appropriate.

Finally, there needs to be a continuous review and improvement cycle to ensure lessons are learned and processes can be optimized on an ongoing basis.

It is clear that in our quest towards achieving the goal of facility integrity excellence, there are a number of complex obstacles to overcome. Although we can never quite achieve excellence, we certainly can drive towards a facility integrity management system that ensures there are no incidents involving breakdowns or degradation of equipment or systems, that maintenance and operations are so well executed that there is minimum effort expended and that the facility availability is almost 100%.

1.4.3 Planning for Facility Integrity Management

Facility integrity management is a continuous assessment process and it should be applied across the life of the facility.

Good integrity performance comes about as a result of good quality design and construction, as well as good operations. It is critical therefore that integrity management activities be initiated at the earliest stage of the facility life cycle, at the conceptual design stage, and continued right through to detailed engineering, construction, commissioning and to the operations phase.

Planning for integrity management of a facility starts in the design phase of the facility life cycle. This process first involves the development of the facility strategic objectives. This is the facility senior management team strategic vision and includes objectives for production, product mix, safety, environment, equipment availability and maintainability strategies. For mature facilities this may include business growth strategies, such as extension of process equipment and systems life, increasing the *mean time between failure* (MTBF) of equipment and extending time between shutdowns.

The planning process moves to defining the integrity organization to sufficiently support the operations, maintenance, and facilities integrity and reliability departments. This process details the numerous organization charts, competence requirements and future training requirements.

During the design phase, the process of planning for integrity management is shown in Figure 1.2.

The next step involves assessment of the risks and evaluation of major hazards. This process details the safe operating parameters (integrity limits) for equipment and also defines the criticality of the facility equipment. This is important so that resources can be assigned sufficiently and availability targets met for facility equipment and systems. A common rule of thumb in the industry is that most of the risk is carried by only a small percentage of equipment and systems. This is a concept illustrated graphically in Figure 1.3. The criticality concept and relationship to the Facility Integrity Excellence Model (FIEM)© shall be discussed in detail in Chapter 4.

It is important that hazards are identified as well as their consequences. This can then be used to make sure controls are in place from prevention to containment and recovery. An effective tool used in the industry is called the *bow-tie model*. This tool introduces layers of protection in order to manage hazards and their consequences. It is used throughout the oil, gas and petrochemical industries for the management of facility process hazards. Figure 1.4 shows a high-level overview of the bow-tie model [1.5].

Bow-ties were first used in the 1970s by Imperial Chemical Industries (ICI) in the UK. Since then, the bow-tie model has been used extensively throughout the oil, gas and petrochemical industries to manage their hazards and consequences.

We will revisit this useful tool during Chapter 6, where we will cover risk assessment in facilities operations.

Once hazards are assessed and risk assessments are carried out, detailed plans are developed and implemented to manage risk. These include specific plans for maintenance, operations, facility integrity and reliability and management of change.

The final step of our planning process for integrity management is to review and learn from our performance. The continuous review cycle includes learning from incidents and equipment failures. We also need to develop a set of proactive key performance indicators (KPIs) in order to develop baseline data, so that we have a basis for improving our performance.

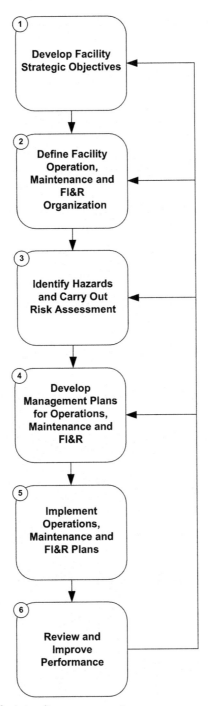

Figure 1.2 *Planning for integrity management.*

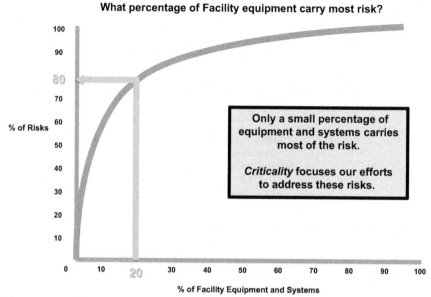

Figure 1.3 *Risk versus percentage of facility equipment.*

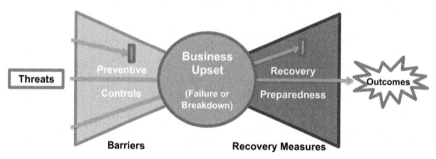

Figure 1.4 *Bow-tie model.*

1.5 INTRODUCTION TO FACILITY INTEGRITY EXCELLENCE MODEL

In its simplest form, a facility integrity management system can be made up of three essential elements. These include a *technical element*, associated with maintenance, facility integrity and reliability; a *personnel element*, which considers the driving force behind integrity management systems, namely the facility organization, its people; and a *support processes element*, which includes a number of core processes and procedures required in order to ensure that the technical and personnel elements are working together with a common goal. Integrity cannot be achieved without these three elements in place.

1.5.1 Holistic Viewpoint

The Facility Integrity Excellence Model recognizes that the successful Integrity programs of this world are founded on the basis that the individual component parts inherently depend on each other in order to be successful and function as a whole. This is because there are significant synergies to be achieved in the numerous overlapping complex processes. These processes all have natural interdependencies and ensuring that these dependencies are integrated and information flow is effective and unconstrained is essential in unlocking the additional value.

This notion is often overlooked or not managed properly in the industry and this can result in high operation and maintenance costs, unreliable equipment and, in some cases, substandard HSE performance. Furthermore, it is noted that, with one of these processes lacking, the whole process of integrity management is vulnerable and there is a higher risk of failure.

The concepts that make up the Facility Integrity Excellence Model are developed with a holistic picture in mind. This is important because integrity management process interdependencies can then be easily identified and properly managed.

The FIEM knits together all of the core and support elements of facility integrity management in order to provide the reader with a comprehensive guide for integrity, maintenance and operations management of petrochemical facilities.

1.5.2 Risk-Centered Culture, the Heart of Facility Integrity Management

As with all successful facility integrity management systems, the heart of the FIEM is its people.

In order to be successful, the integrity organization must have the right beliefs, the right mindset and way of working, or *culture*. Effective integrity management of a high hazard facility requires integrity processes to be embedded at all levels of the organization, top- down and bottom-up, from the policy to the strategy and finally to the work procedures and standards that operate within the company.

"A change sticks when it seeps into the bloodstream of the corporate body" [1.6]. Until new behaviors are rooted into social norms and shared values, they are subject to degradation as soon as the pressure for the change is removed. It is the culture of the organization that is the target for change in order for the implementation of the excellence model to be effective.

Because this is so important, we shall explore in detail how to go about implementation of major change, such as a new integrity management system, in Chapter 11.

Risk-centered culture (RCC) is the term given to the heart of the excellence model, which represents the organizational culture. Culture is a key component of the concept of RCC, which considers the way the facility organization's employees behave. Organizational discipline underpins all effective risk management systems. When the major oil and gas incidents of recent times are analyzed, such as Piper Alpha, it is clear that poor organizational discipline has led to inadequate risk management resulting in a major incident that could have easily been prevented. Risk-centered culture will be presented in detail in Chapter 3.

1.5.3 Facility Integrity Excellence Model Representation

The FIEM is represented in Figure 1.5. We can see that in the center of the model we find risk-centered culture. The intention here is to highlight the importance of the organizational culture being in place and properly functioning in order for a facility integrity management system to be effective.

1.5.4 The Three Essential Elements of Facility Integrity Management

We have already discussed that facility integrity management needs to consider three essential elements which are fundamental to effective integrity management. Within the FIEM these are: *facility*, which addresses the technical aspects; *personnel*, which focuses on the resourcing and competence of the integrity organization; and *supporting processes*, which address the key processes required to support the integrity management system.

The three essential elements are represented as segments arranged around RCC (in the center of the model) in Figure 1.5. This symbolizes their dependency on ensuring the foundation is in place – a healthy organizational culture. These essential elements or segments are further detailed in a series of work flows which will be discussed in the following chapters.

To outline the essential elements, we will first look at the technical element. 'The Facility' segment of the excellence model is intended to provide a detailed understanding of the key processes that ensure the safe and effective operation and maintenance of the facility from a technical standpoint. This segment of the excellence model comprises three aspects. First is

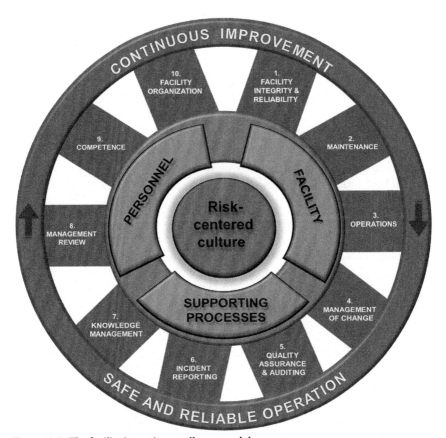

Figure 1.5 *The facility integrity excellence model.*

maintenance, which is concerned with quickly correcting failures, minimizing equipment and systems downtime, while balancing the associated resources and cost. Second is *facility integrity and reliability*, which applies risk-based methods to ensure the integrity and reliability of critical equipment and safety systems, describing the ability of equipment or systems to function under stated conditions for a specified period of time. It also focuses on ensuring that equipment and systems are properly designed, installed in accordance with specifications, and remain fit for use until retired. The third aspect is *operations*, which is concerned with ensuring the facility is operating on demand, efficiently and within its integrity limits.

There are a number of fundamental processes that support facility integrity; these are represented in the model under the second segment, 'Supporting Processes'. The first process within the supporting process segment is *management of knowledge*, which focuses on capturing and learning from performance data from each of the facility segment processes and ensuring this is put to good use. For example, a minor reoccurring equipment failure over time will consume a considerable amount of resources. If failure data in this scenario is collected and analyzed, the root cause of the equipment failure may be identified and resolved, which eliminates the reoccurring failure and wasted ongoing resource allocation. Management of knowledge is very important but also making sure that relevant information is collected and that it is used correctly is equally as important. The Supporting Processes segment also includes processes for *incident reporting, quality assurance and auditing,* and *management of change.* These processes are detailed in Chapter 6.

'Personnel' segment of the model underpins the Facility and Supporting Processes segments in that it drives towards ensuring an appropriately resourced and competent organization is in place to support the facility. In the absence of a competent integrity organization, all efforts to deliver facility integrity will be ineffective.

1.5.5 "What Gets Measured Gets Done"

Continuous improvement is a fundamental quality of the FIEM. It is absolutely essential to measure the performance of the facility in order to address shortcomings in equipment and systems performance.

In addition it is important to measure performance of the integrity processes, not just the performance of facility equipment and systems. This is in order to enable the improvement of the integrity management system as a whole. FIEM assigns key performance indicators to its processes in order to manage and improve their performance.

It is important that a comprehensive mix of KPI data is measured and presented in an easy-to-understand dashboard in order to "tell the full story" rather than an incomplete version. This will include a balance of information that relates to historic performance (reactive KPIs) and information that can give indication of future performance (proactive KPIs).

FIEM also takes into account hierarchy of reporting, which aims to present information as it is required at the different management levels in the integrity organization. For example, at senior management level, information about total recordable incident rate (TRIR) relating to safety performance and profit and loss will be required. Within the maintenance department typical information that may be required to improve performance will be about equipment performance, such as *mean time between failure* (MTBF). The Continuous Improvement element of the FIEM will be presented in detail in Chapter 9.

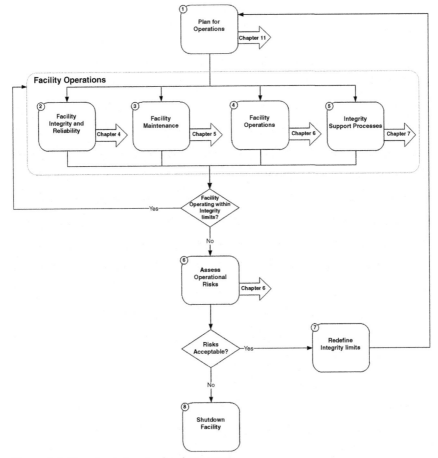

Figure 1.6 *Flowchart showing book structure.*

1.6 DESIGN CONSIDERATIONS FOR THIS BOOK

This book has been designed and structured in a unique way, for ease of understanding and application of the concepts that describe the Facility Integrity Excellence Model, which applies to both existing and new facilities. Figure 1.6 shows a flow chart with the core integrity work flow process identified. Each of the core work process is described in detail in the chapters highlighted, which makes up the structure of the book.

The following premises were used during the development of this book to ensure this principle is adopted:

1. Holistic approach to facility integrity management with workflow processes detailing the key concepts and how they function together as a system.

2. Solid foundation of knowledge and information.
3. Practical application of the key concepts, in terms of the author's viewpoint rather than a theoretical account, which is often difficult to digest and apply to real production facility scenarios.
4. Quick and easy to read and page through to relevant theory and applications.
5. Aimed at a wide-ranging audience including graduates, young engineers and managers.

This book also considers the softer side of the organization and presents effective techniques for the implementation of a new facility integrity management system in existing facility organizations. Organizational change management is a difficult process to get right especially due to the fact that effective integrity management has many stakeholders throughout the organization. FIEM presents effective tried and tested techniques for implementing changes of this nature in large complex organizations. There are real financial gains to be realized through implementation of the concepts presented in this book.

1.7 CHAPTER-BY-CHAPTER SYNOPSIS

1. Introduction. Chapter 1 sets the scene for facility integrity management. It discusses how, more often than not, there are gaps in the industry's approach to integrity management. It introduces the concept of an integrity excellence model that can address some of the shortcomings, focus on improving facility performance and save on costs.

2. Facility Integrity Excellence Model. Chapter 2 explores the Facility Integrity Excellence Model in detail, explaining how each of the elements of the model work and fit together and the resulting synergies can be realized. The chapter presents the case for integrity management by reviewing a case study that explains what can happen when integrity goes wrong and how integrity management systems are of paramount importance in the effective day-to-day operation of an oil, gas or petrochemical facility.

3. Risk–Centered Culture (RCC). The heart of the facility integrity excellence model is its people. The concept of risk-centered culture is presented in Chapter 3, which describes the collective beliefs of the integrity organization. RCC ensures that these beliefs are aligned to achieving integrity excellence and are engrained into the organization from the shop floor level to the top management.

4. Facility Integrity and Reliability. The key concepts of facility integrity and reliability (FI&R) are presented in detail in Chapter 4. FI&R management is the means of ensuring that the people, systems, processes and resources, which deliver integrity, are in place, in use and fit for purpose over the whole life cycle of the facility. The chapter explains how FI&R can influence the profit and loss of the facility. The chapter explores failure of equipment and systems from first principles, and presents sound root cause analysis techniques that have been tried and tested in the field. The key concept of facilities equipment criticality is presented and the chapter shows how many of the excellence model concepts are integrated within this concept. Finally the chapter explores the shift from a reactive to a proactive facility management strategy.

5. Maintenance Management. This chapter introduces a number of common failures in facilities maintenance management that are well known throughout the industry and discusses how these failings can be overcome through the concepts introduced in the excellence model. The notion of reactive versus proactive maintenance is explored and the key maintenance strategies and maintenance management principles are presented by introducing the excellence model maintenance element of the facilities segment.

6. Integrity Operations Management. Integrity operations management involves monitoring operations performance and responding to different scenarios that come about. It defines the concept of operational variances which occur when the facility is not operating as planned. It also introduces integrity limits in order to control performance and ensures safe operation. The chapter then explores unit monitoring principles and details the techniques for assessment of equipment condition, early warning of incipient equipment failures and equipment degradation. It proposes tried and tested methods for equipment surveillance, reporting and prompt response in order to minimize equipment outage.

7. Supporting Processes. There are a number of supporting processes that are required in order to ensure facility integrity management is effective and sustained. These supporting processes are detailed in Chapter 7 and include management of knowledge, management of change, incident investigation and management reviews.

8. The Integrity Organization. The effectiveness and success of any new initiative lies in its people. The people behind the scenes, operating and running the organization, are by far the most valuable resource the company has. The organization is a critical component of the facility integrity excellence model and is presented in Chapter 8.

9. Continuous Improvement. Facility integrity excellence model continuous improvement focuses on ongoing review and improvements to facility performance as well as facility integrity management processes. It details key performance indicators in a series of dashboards that are presented in a hierarchical reporting system in order to convey the right information to the right level in the organization.

10. Implementation of FIEM. Organizations are about people, their development, enhancing their performance and building the organization on their performance. FIEM considers the difficulties when introducing change into organizations. No matter how welcoming an organization is to change, it will still face a degree of employee resistance to change. This chapter presents tried and tested methods for effectively managing change in organizations and ensuring implementation of facility integrity management systems can be achieved successfully.

11. Facilities Integrity and Strategy. Chapter 11 explores integrity management strategy, which includes the development of a strategy framework. The framework includes development of an integrity policy and strategy along with specific requirements for the systems and personnel which encompass all of the necessary roles and responsibilities for integrity management.

CHAPTER 2

Facility Integrity Excellence Model

Contents

2.1 DEVELOPMENT OF THE FACILITY INTEGRITY EXCELLENCE MODEL

The Facility Integrity Excellence Model (FIEM) provides a solid foundation for facility personnel to work towards achieving excellence in the integrity performance of their facility. The FIEM has been developed using a set of philosophies that ensure it is robust and that the basis is founded on best practices with consideration for practical application on a facility. The FIEM illustration is shown in Figure 1.5. We shall now focus for a moment on some of the key philosophies that have been applied to the FIEM in order to meet the design considerations as described in section 1.6 in Chapter 1.

In Chapter 1 we represented the approach to "planning for integrity management" as a process model, as shown in Figure 1.2. This approach enables us to break down each process into digestible pieces in order to help us understand them more clearly. The key concepts that make up the Facility Integrity Excellence Model follow this same philosophy.

Facility Integrity Management
http://dx.doi.org/10.1016/B978-0-12-801764-7.00002-4

2.2 PROCESS MODELING

Process models can be used to provide an accurate representation of a reality based on a set of key principles. They can offer significant advantages if deployed correctly. They can be used to prescribe how things should be done in order to do them well and achieve a desired state. They can also help to provide an insightful level of clarity around complicated processes and environments. Models are an anticipation of what a process should look like; however, the actual performance of the process will be determined by how well they are implemented. Implementation aspects of the Facility Integrity Excellence Model are presented in Chapter 10.

Management of facility integrity is a complicated topic with many aspects to consider. Representing the management of facility integrity as a model provides us with an easy-to-use tool that can help us to view integrity management at all levels. At the highest level it helps us to understand how all of the key elements fit together and operate as a whole. At the lowest or most detailed level it helps us understand how the key processes work and operate together.

This pragmatic attribute can also help us to isolate the key processes and break them down to a sufficient level of detail in order to understand the most critical elements and to ensure these elements are given the priority attention they require in order to ensure the whole process operates effectively.

Business process models also have another important advantage in that they can be represented as a series of layers. This allows a process to be broken down into further detail which adds the much-needed definition.

The different layers can represent the facility integrity management processes at various levels within the organization. At the highest layer, *strategic processes* are aimed at defining and developing the strategy of the organization in order to achieve better performance. They are a process of continuous improvement and involve measuring historic performance, addressing gaps and optimizing processes. They must also take into account the external environment which may have an impact on performance. An illustration of business process model layers can be seen in Figure 2.1. As we break down each strategic process, we can delve into the detail as needed to sufficiently describe each process.

If we refer to Chapter 1, we described a workflow process in Figure 1.2 of planning for integrity management. This is a strategic level process describing how to approach facility integrity management from a strategic perspective, which will be explored in detail in Chapter 11. If we expand on

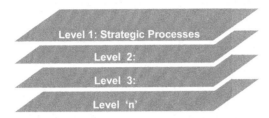

Figure 2.1 *Abstraction of a process model.*

the idea of abstraction of a process model, we see that describing processes by using layers can add much needed definition and clarity.

It is well known and understood in the industry that processes have three components, comprising *inputs*, a *process* and an *output*. Fundamentally, workflow processes convert inputs into outputs by doing work. This may be converting refinery raw materials such as crude oil via a refining process into a salable product mix, which may be made up of diesel, petrol and gas. Or in the context of facility integrity management, we may be improving equipment reliability by making modifications to equipment in order to enhance performance based on historic performance data.

Work flow processes are important because they enable us to understand how our internal processes and systems function. This is crucial in order to understand how we can make improvements. We need to know how we are performing today in order to improve our performance in the future. Improvements may result in lowering operating overhead costs, increasing throughput, pushing out time between shutdowns, de-bottlenecking, minimizing waste, etc. We can therefore with relative ease apply a measurement system not only to measure the results of the facility integrity processes but also to measure internal performance of our integrity processes themselves.

Ultimately, we are driving towards an answer to the following questions: How is the facility integrity department performing?

Can we improve the way the facility integrity department executes its work?

Are there new processes that we can introduce to further improve facility integrity performance?

The principle of workflow processes is applied to provide the much-needed clarity and definition of our performance today. This gives us a solid basis to:

- Establish our current way of working and understand why we do things this way;
- Provide a basis for measuring our performance;

- Show the flow of information to and from the various activities and how it is converted from inputs into outputs;
- Identify any key bottlenecks that may result in inefficient or ineffective processes.

The Facility Integrity Excellence Model is made up of a series of workflow processes. Certain workflow process outputs may be in the form of documents, or information that provides inputs to other processes. In order to provide structure to each of the workflow processes, a well-known and effective quality control system is applied.

2.3 THE SHEWHART CYCLE

The Shewhart Cycle was developed by Edward Deming [2.1], who was considered by many as the father of modern quality control. The Shewhart Cycle was based on a scientific method thought to originate from the work of Francis Bacon. The basis was written as "hypothesis" – "experiment" – "evaluation" or "plan" – "do" – "check". The final step of the cycle, "act," was added by Shewhart in order to take action based on the conclusions of the evaluation. The steps of the cycle then became "plan" – "do" – "check" – "act" or PDCA (Figure 2.2).

One of the fundamental philosophies of PDCA is the concept of continuous improvement. Once the output of a PDCA process is "checked" there is a response, which is based on data, in the "act" step, driving us one step closer to improving our process. All good integrity processes should contain all of the elements of the PDCA cycle. PDCA is a continuous improvement cycle which involves iteration in order to constantly drive to facility integrity "excellence."

Figure 2.2 *The Shewhart Cycle.*

The Facility Integrity Excellence Model workflows are based on the Shewhart PDCA cycle for several important reasons. Firstly, PDCA involves clear distinction between the end of one step and the start of the next. This is important because if workflow is unclear, then by nature there is an element of ambiguity. This may ultimately be a source of error in the FIEM which needs to be eliminated.

Secondly, PDCA ensures that all FIEM workflows are driving towards operational excellence, which is based on the principle of continuous improvement.

PDCA is based on a scientific method, it is problem solving by nature. By deploying PDCA in workflow processes, the FIEM organization is conditioned to think with a problem-solving mentality, constantly challenging themselves and the integrity processes in order to better understand them and improve on current performance. This contributes towards development of a risk-centered culture (RCC) which is explored in detail in Chapter 3.

The key steps of the Shewhart Cycle are explained as follows. In order to put this key philosophy into the context of the FIEM, it has been referenced to Figure 1.2 presented in Chapter 1 (planning for integrity management).

2.3.1 PLAN

The PLAN step of the Shewhart Cycle establishes the objectives up front. These are required in order to ensure there is a benchmark for comparison against the results of the process. By establishing a benchmark it is possible to ensure that each process can be continually improved. In the context of planning for integrity management, which is shown in Figure 2.3, this activity focuses on developing a set of strategic objectives which include providing a strategic direction along with resources needed to execute them.

2.3.2 DO

The DO step refers to the implementation phase of each process. It is the work done that forms part of each process. In the case of planning for integrity management, this step addresses the implementation phase of the plans made in the previous step.

2.3.3 CHECK

The CHECK step measures the performance of each process from the data collected in the DO step and compares it against the benchmark results from the PLAN step. Deviations from the plan and completeness of each process

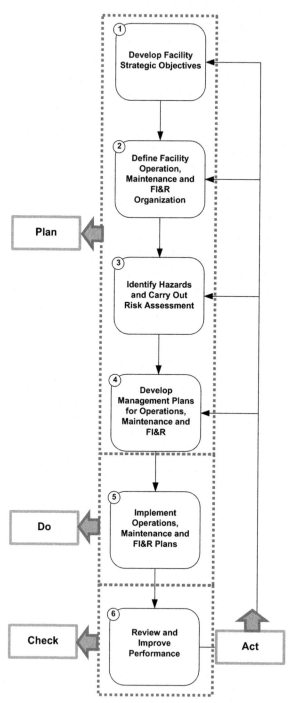

Figure 2.3 *The Shewhart Cycle applied to planning for integrity management.*

are measured on an ongoing basis in order to develop a trend. Trending is important so we can understand how we have performed and what corrective action we need to take.

2.3.4 ACT

The ACT step involves addressing any deviations in our process in order to continually improve our performance. This step takes the form of an analysis of the deviations in order to understand their root causes. A root cause analysis may be performed during equipment failure. Corrective action needs to address the root cause and not just focus on reinstating the equipment back into service as quickly as possible. This would result in another failure in the future. We will consider root cause analysis in detail in Chapter 4.

The ACT step also provides a mechanism for continuous improvement, especially when we want to assess whether our strategic planning for managing the integrity of our facility has been successful.

2.4 BEST PRACTICES

Best practices are used to set a benchmark, maintain quality and ensure the right actions are taken to deliver excellence in integrity management. The term "best practices" is often used as a buzzword and these practices often evolve over time to better follow the development of new techniques and tools for the execution of activities. However, in this instance, the term "best practice" is used to describe an industry norm, a set of activities to achieve an outcome that has been tried and tested in the industry.

The workflow processes developed as part of the FIEM are developed with best practices in mind and tailored in order to factor in practicalities of day-to-day operation along with the operational constraints that facility personnel experience on a daily basis. These may include financial, operational and resource-based constraints. Further useful templates based on "best practice" in order to standardize workflows have been developed for the reader for certain integrity processes in the following chapters.

2.5 COMMON LANGUAGE

In order to be consistent and avoid misunderstanding, a "common language" philosophy has been developed and used throughout this book.

It is pertinent to first be clear on the key definitions of the core workflow processes that make up the Excellence model. The model views integrity from several bases: a technical basis, i.e. the facility, considering best practices for maintenance, reliability and operations; a personnel basis, which looks at best practices for personnel organizations, addressing management and development of staff within integrity organizations; and also the supporting processes associated with integrity management, such as management of change, incident reporting and knowledge management.

The following list of key words and phrases and their corresponding definitions are used throughout this book:

Facility: A technical basis considering best practices for maintenance, reliability and operations.

Personnel: A basis that looks at best practices for personnel organisations, staff management and development within integrity organizations.

Support processes: A set of supporting processes associated with integrity management, such as management of change, incident reporting and knowledge management.

Maintenance: Concerned with quickly correcting failures which are driven by a natural law of system changes. The ultimate aim of maintenance is minimizing maintenance cost and downtime. In most cases at a facility, maintenance is carried out in response to equipment failures. There are few incidents where planned maintenance is carried out due to the nature of the continuous operation of the facility other than planned unit shutdowns.

Reliability: Concerned with avoiding failures of equipment and processes by proper design and careful operation of the equipment by trained personnel in a specified environment for a given time interval. The ultimate aim of reliability is a failure-free environment. Close monitoring of equipment and plant performance offers the opportunity to avoid failure mechanisms and to detect early signs of deterioration or equipment distress that can be remedied before failure. A failure-free environment allows the plant to be utilized more effectively and ultimately increase production output. Essentially, reliability is all about making money. Typically, reliability is associated with rotating equipment on oil and gas or petrochemical Facilities.

Integrity: A facility has integrity if it operates as designed with all risks as low as reasonably practicable. Typically integrity is associated with static equipment on oil and gas or petrochemical Facilities.

Workflow process: A representation of an integrity process by a series of activities that flow from start to end. The workflow process includes inputs and outputs and shows the flow of information.

Workflow process hierarchy: Individual activities within a workflow process can be further developed into more detailed processes.

Lower level process: A workflow process that has been developed from a high-level process to show further detail as a part of the workflow process hierarchy.

Uptime: The proportion of time facility equipment and systems can operate at its maximum demonstrated capacity while producing quality product.

Downtime: The proportion of time facility equipment and systems are unable to provide or perform their primary function.

Availability: The proportion of time a system is in a functioning condition, available for its designated tasks.

2.6 THE IMPORTANCE OF INTEGRITY MANAGEMENT

Now that we have discussed the key design considerations that have shaped the Facility Integrity Excellence Model, we shall now look as a case study that illustrates the necessity to ensure that a robust integrity management is implemented well within the facility organization.

2.6.1 Case Study – ConocoPhillips, Humber Refinery

The ConocoPhillips Humber Refinery can process nearly 250,000 barrels of oil a day into a range of products, from low-sulphur petrol and diesel to liquefied petroleum gas (LPG), heating oil and industrial feed stocks such as propylene [2.2]. The refinery is a Major Accident Hazard site as defined under the Control of Major Accident Hazard Regulations (COMAH) 1999.

On 16 April 2001 a huge fire and explosion occurred on the Humber Refinery. The incident had the potential to cause fatal injury and environmental impact. It was extremely lucky that on that particular day there were no serious injuries. In terms of damage, however, it was extensive, as much of the refinery was destroyed and damage to local property was evident at over 1 km away from the blast.

The incident occurred on the Saturated Gas Plant (SGP), where an overhead pipe carrying flammable gas under high pressure ruptured. As soon as the pipe ruptured, a huge gas cloud formed containing a mixture of hydrocarbons. A short time later, the gas cloud found an ignition source which resulted in a considerable explosion and fire.

In total there was approximately 180 metric tonnes of flammable liquids and gases released. The incident also resulted in over half a tonne of

Figure 2.4 *Explosion at the ConocoPhillips Humber Refinery, 16 April 2001.*

an extremely toxic gas, hydrogen sulfide, released. The SGP suffered catastrophic damage due to the sheer force of the explosion and the resultant fire caused further damage (see Figure 2.4, which shows a photo of the aftermath of the explosion).

The incident had the potential to cause fatal injury and environmental impact, although no serious injury occurred and there was only short-term impact on the environment. There was, however, major damage to the refinery and to nearby properties.

2.6.1.1 Root Cause

Under the COMAH regulations, which implements the Seveso II Directive in Great Britain [2.3] it was required that the incident has to be investigated by the joint inspection and enforcement body, known as the Competent Authority. This comprises the Health and Safety Executive (HSE) and the Environment Agency (EA) [2.4] in England and Wales. The Competent Authority is required by the Seveso II Directive to carry out regulatory activity including inspections in order to ensure that company operations are being conducted in accordance with legislative requirements. This is so that root causes of the incident can be ascertained and corrective action taken, so that the industry can learn from the incident.

The investigation managed to locate the source of the explosion, which came from the elbow of an overhead gas pipe on the refinery SGP. It was noted that soon after the commissioning of the SGP, salts or hydrates began to accumulate in the heat exchangers downstream, causing fouling problems. In response to this problem, an existing vent point on the pipe was modified to include the injection of water into the line. The idea was that the water injected into the pipe would dissolve the salts to address the fouling problems.

It was discovered that, over time, the water injection into the vent impinged the pipe elbow in a corrosion/erosion degradation mechanism and ultimately ruptured the elbow. On investigation it was noted that the overhead gas pipe elbow wall thickness had reduced to such an extent that the wall could no longer withstand the internal pressure within it. Furthermore, it was noted by the investigation team that the pipe was internally coated with black iron sulfide (passivation layer), which provided a protective corrosion layer.

Unfortunately, with the newly introduced water injection, the passivation layer was washed away over time by the water injection. The investigation team concluded that the failure mechanism and resulting loss of containment was down to erosion–corrosion of the pipe due to the uncontrolled installation of the water injection. There was no assessment of the impact of the installation of the water injection modification. There was also no consideration for water dispersal such as an injection quill, and as such the water entered the pipe as a free jet which would act to accelerate the erosion–corrosion effect. The investigation team noted that the modification had the hallmarks of a "quick fix."

2.6.1.2 The Barriers that Broke Down

Within a competent integrity organization we put barriers in place in order to provide safeguards and controls to prevent incidents. Barrier analysis is often compared to "Swiss cheese" which is stacked together side by side. Each barrier is represented by one slice of cheese. The holes in each slice of cheese represent weaknesses in the barriers. When a hazard is present the barriers act to prevent it from propagating into an incident. Incidents occur when the holes in each slice of cheese line up, which permits the propagation of the hazard, which ultimately results in an

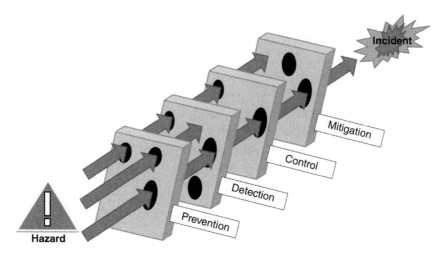

Figure 2.5 *Swiss cheese model – barrier analysis.*

incident. The hazard picks up momentum as each barrier is overcome. As we apply the barrier analysis model, it is apparent that weaknesses within each barrier tend to be present in some shape or form. In continuation of the Swiss cheese analogy, we recognize that the "holes" are continually under a state of change: for example, differences in competence of operations and maintenance staff, equipment wear and performance deterioration, bypasses of operation and maintenance procedures, along with many others.

Our aim in this regard is to identify the "holes" or weaknesses and ensure that they are as small as possible. Additionally, we can consider additional Swiss cheese slices or barriers. This in effect reduces the chances of the holes lining up and therefore the possibility of a hazard propagating through each barrier and resulting in an incident.

We can apply a simple barrier analysis to this case study to illustrate the main elements that led to the incident on the Humber Refinery, illustrated in Figure 2.5.

Prevention

The first barrier in this model is prevention. How can we prevent a failure occurring in the first place?

In this case, it comes down to ensuring the design is correct and having appropriate processes in place. This would have been captured in a management of change process. However, this was not the case for the water injection modification to the SGP. The design did not account for the environment it was to be installed into, which ultimately led to the failure. We could have designed out the failure in the first place by proper design and this would have been facilitated by proper management of change.

Detection

The second barrier is detection. How can we pick up the onset of a failure which is about to occur? What tools do we need to have in place to detect the early warning signs of a failure?

Monitoring and inspection are critical elements of the FIEM. We need to be able to assess risk as an organization and to appropriately address it. In the case of the Humber Refinery, a degradation mechanism that would end up in a huge explosion was not detected because the facility organization was unaware of associated risks of an uncontrolled modification and had not put a barrier in place to detect it.

Control

The penultimate barrier is control of the failure. We are now moving closer to ultimate failure; we failed to design out the failure and we failed to detect it in time for an early effort to arrest it, how can we now put controls in place to prevent the ultimate failure?

In this case we could have introduced a new control mechanism to safeguard against this eventuality. An improvement to the existing facility emergency shutdown system may have provided early warning in terms of gas detection or emergency shutdown.

Mitigation

The final barrier is our last hope, to mitigate the failure. This barrier is aimed to protect as much as possible our facility personnel, the facility and the environment. What mitigation plans could we have put in place to minimize the impact of the failure?

The reliance was on the blast-resistant control room, which for all intents and purposes appeared to be effective in protecting the facility workforce.

2.6.1.3 Outcome of the HSE Inquiry

There were a number of key outcomes from the inquiry by the Health and Safety Executive. Interestingly, many of these outcomes provide the building blocks for an effective and robust facility integrity management system such as FIEM.

First, the design and installation of the water injection point was not subject to a robust risk assessment nor did it follow a management of change assessment. The investigation team noted that if there had been a risk assessment done, the high risk of corrosion could have been identified and action taken. In addition, there was no risk assessment made regarding the change of the original intent of the vent pipe to use as a water injection point over the lifetime of the plant.

The inquiry noted that effective management of change systems, which consider both plant and process modifications, are critical in order to prevent major accidents. Management of change is a key workflow process that forms a part of FIEM.

Secondly, the inquiry noted that the process for the inspection of piping was not effective and was not implemented well. The system that was used fell short of industry best practice and failed to use knowledge and experience from other sections of the facility, which would have helped identify the SGP as an area of concern. Inspection systems are an essential part of a facility integrity management system. It is imperative to identify where inspection should take place on facility processes and equipment but also it is equally as important to identify the inspection intervals so that it is possible to ascertain the degradation rate of equipment and processes. Within the FIEM there is a risk-based approach that assesses the priority areas so that the facilities workforce can focus their efforts much more effectively.

Finally, the inquiry pinpointed failings in communication that led to the incident. The investigation team noted that ultimately the modification to introduce the water injection point was only communicated to the facility operations team and was not discussed or understood with the wider facility teams.

As with any integrity management system, effective communication is a critical element. Encouragement of sharing information between the numerous facility teams and personnel is imperative in order to prevent major incidents.

We are aware of the major oil and gas disasters over the last few decades because these are the ones that have received a good dose of publicity, since their exposure was too great to contain. The number of integrity incidents that have caused serious injury and facility damage and that we are not aware of because they did not receive press coverage to any great extent is many times more. One of the basic principles of integrity management programs and the FIEM is to prevent such occurrences.

2.7 THE ROAD TO FACILITY INTEGRITY EXCELLENCE

We can see in the case study on the Humber Refinery that all of the barriers broke down leading to a significant incident. It was not understood at the time that there was such a catastrophic failure mechanism in play, which could have been easily prevented, detected, controlled and mitigated.

The road to facility integrity management (Figure 2.6) requires a dedicated effort by the whole facility organization. There are certain infrastructure requirements that need to be developed and implemented but ultimately it is the mindset of the facility organization that needs to think differently and approach integrity from a proactive viewpoint.

We may start with a scenario whereby we feel that we are constantly responding to equipment failures and equipment malperformance. This is commonly referred to as "reactive mode." Operating a facility in reactive

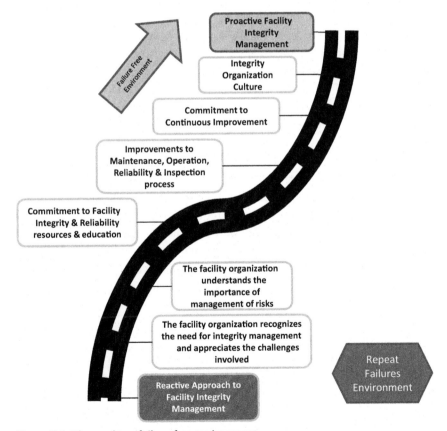

Figure 2.6 *The road to a failure-free environment.*

mode is like waiting for the symptoms of a serious illness to be severe enough to disrupt your normal routine. It is expensive, disruptive, and ineffective; it is the oil and gas facility equivalent of using a band-aid or plaster on a serious injury.

So how can we move from an environment where we see repeat equipment failures to an environment where we are failure free?

We may be familiar with the term "fighting fires," which results in moving into a reactive mode. This situation is self-perpetuating and can consume inordinate amounts of time and effort for many of the facility organization teams. We approach an equipment failure with the mindset of reinstating it in the quickest possible manner but without really understanding why it failed. This will inevitably lead to repeat failures of the same nature. We leave the failure site with little or no information, which results in us having to revisit the same failure site to address the same failure at an unknown time in the future.

We may come to a point whereby the facility organization appreciates the challenges it has and recognizes that there is a need to change in order to escape fighting fires or the "repeat failures environment" (Figure 2.6).

The next stop along the road to facility integrity excellence is to start to consider changing the way we think and to start to consider our facility in terms of risk. This enables us to assess the severity and consequences of the events taking place, which in turn allows us to apply our efforts in a proportionate manner in order to effectively manage risk without wasting resources. We will need to educate our organization and may consider training programs and onboarding additional resources to fill skills gaps.

Moving along the journey, we will review our existing facility processes and systems within our organization departments and align them to meet the key principles behind a facility integrity program or FIEM. It is necessary to consider the importance of a standalone reliability and integrity group to work alongside operations and maintenance teams. This is a key principle behind the FIEM; however, it is imperative that this group integrates with operations and maintenance in order to be effective.

We are now starting to see a return on our investment; we have a facility organization that understands the value of performing risk assessments on the facility and prioritizing invaluable resources. We have aligned our facility processes and systems with integrity management principles. We have a new reliability and integrity team that complements our existing facility organization teams tasked with ensuring our facility is healthy and available when needed.

Our culture is starting to change and we are starting to adopt the traits of a proactive facility organization driving towards a failure-free environment.

2.8 FACILITY INTEGRITY EXCELLENCE MODEL WORKFLOW

A facility has integrity if it operates as designed with all risks as low as reasonably practicable. Maintaining facility integrity requires a focus on people, equipment, processes and procedures. Facility integrity management is a continuous, proactive process applied throughout the facility life cycle. There are many definitions of a facility integrity management system; however, common to all is this definition of a facility that has technical integrity: "A facility that meets its intended purpose where there is no probable risk of failure that may result in safety of personnel, environment or facility damage. In the event of deviations from the intended purpose, these have been identified and risk assessed and mitigation actions are in place to ensure continued safe, efficient operation."

A key attribute of the FIEM is that individual elements are represented in a series of workflow processes. The reason behind this is in order to break each of the elements down into first principles. This facilitates a clear and unequivocal understanding behind each of the processes, which includes their inputs and outputs.

The FIEM does this by employing a technique called business process reengineering. BPR is a business management strategy, which focuses on the analysis and design of workflows within an organization. It is a well-known and effective tool that was originally established in the 1990s.

BPR allows the existing core facility integrity process to be dissected, evaluated from first principles and reconstructed in order to represent the most effective way to manage integrity of a facility. The reconstructed processes take the form of a series of workflow processes that cover the central concepts of the subject matter. Each of the workflow processes are developed to a sufficient level of detail, including the critical information flows as inputs and outputs to each process. The processes also reference material in the subsequent chapters of this book for ease of reference. This concept affords the reader a holistic approach to and complete understanding of the key integrity, reliability and maintenance management principles at their fingertips, with the option of quickly referencing the key subject matter as required.

At the highest level we can represent facility integrity management in a process workflow following the business process reengineering (BPR) technique. This shows the core processes and their association in the overall workflow. In Chapter 1, we represented our planning workflow process for integrity, which is the first step in our Integrity journey (section 1.4.2, Figure 1.2, Management of integrity during the design phase of a facility).

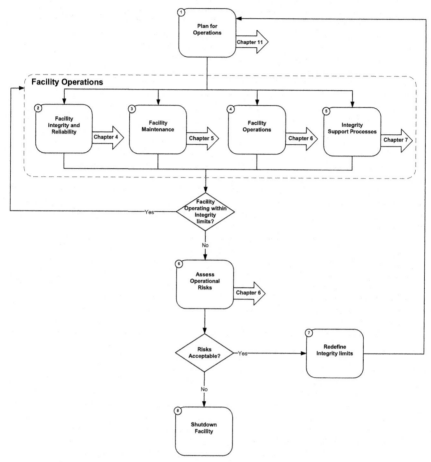

Figure 2.7 *Facility integrity management workflow.*

This involves strategic activities, including the development of an operational strategy which encompasses production targets, equipment uptime targets, inspection and maintenance regimes, planned shutdown targets, and HSE compliance work as well as identification and mitigation of hazards.

We can now move on to management of integrity as we are operating the facility as shown in Figure 2.7. It is noted that citations to the relevant chapters in this book are made beside each of the processes in the workflow shown, which are designed in order to simplify referencing. Each process is presented in detail in the following chapters.

The first step of the process is planning for integrity, which we have detailed in Chapter 1 (Figure 1.2). We shall look at the strategic elements of this process further when we visit integrity strategy in Chapter 11.

The second step of the process is facility operations, which represents the operational phase of the facility. This is made up of a series of detailed workflow processes which include maintenance, integrity and reliability, and operations and support processes.

While the facility is in operation, the performance of the facility equipment and systems is monitored through various means, which may include a suite of inspection and condition monitoring techniques and activities. Under normal conditions the facility operates within its designed integrity limits. The facility performance trends are assessed continually. If an integrity limit is breached there may be a number of resulting occurrences and actions that can be performed by the integrity organization in order to control the situation.

Associated risks must be assessed in order to determine whether the situation is of concern enough to warrant a shutdown. An example of this may be a situation that presents a safety risk. If the risk identified can be managed during operation, an assessment is to be carried out to fully analyze the risk and mitigating actions in order to sufficiently redefine the integrity limits. This may be performed through a management of change process.

2.8.1 FIEM Key Systems

The aim is for facilities that deploy an FIEM or similar models to operate in the top quartile in terms of facility uptime and operating costs when benchmarked with the competition. FIEM facilities aim to be "best-in-class". The FIEM sets the direction through the key processes that make up the model. It intends to maximize the availability of the facility, at optimize cost, with no harm to people or the environment and while safeguarding integrity.

The FIEM is essentially constructed based on three key systems:

1. **Management Systems:** These systems are critical since they provide much needed structure to the integrity effort including policies, management arrangements, roles and responsibilities for the integrity teams, competency assessments, workflow processes including the supporting processes segment of the FIEM model, such as management of change procedures, incident reporting and audit processes.

2. **Engineering Systems:** These systems include methodologies used to define the necessary integrity management processes, such as risk-based methods (RBI and RCM), design codes and standards, and assessment of variances in equipment and systems performance.

3. Facility knowledge management and document control systems: Management of knowledge includes maintenance, inspection, condition monitoring, equipment and systems records and performance history; testing schedules; tracking of variances; records of engineering, procurement and construction of the facility; and records of modifications to the facility, among others.

As well as being a database of knowledge and information, the FIEM needs to be able to facilitate the flow of information as required. Depending on the complexity of the type of oil and gas or petrochemical facility, the FIEM needs to be structured in such a way as to ensure that its interfaces are sufficiently managed. For example, the condition monitoring group within the maintenance department, the static inspection group within the FI&R department, and the operator visual inspections team within the operations department all need to be clearly defined and integrated together. There needs to be certainty of information transmitted and received between the different FIEM functions in order to ensure the excellence model is effective.

If we revisit the FIEM representation as presented in Figure 1.5, shown here in Figure 1.5, we can see there are 10 essential elements, each belonging to one of the three core segments: Facility, Supporting Processes and Personnel. When we look at the interrelation between each of the elements at this point, we can see that a huge amount of integration and information flow is required between the elements.

It is imperative that the information flow between each element is clearly understood and flows smoothly. Interrelations between the different elements may be in the form of interdepartmental meetings, updating logs with information from other elements, joint reviews of equipment performance or failures among many others. Knowledge management will be presented in detail in Chapter 7, covering integrity supporting processes along with incident reporting, quality assurance and auditing, and management of change.

CHAPTER 3

Risk-Centered Culture

Contents

3.1 INTRODUCTION

Managing risk is part of everyday life. If we evade all risk we would see no progress; on the other hand, with a high tolerance for risk we would end up with another Piper Alpha Disaster or explosion on the Humber Refinery.

Facility managers have key business objectives that are set and that need to be achieved in order to meet their operational targets. The uncertainty of achieving their objectives may be categorized as a risk. The management of risk is the systematic process of identifying, assessing and addressing risk, so that we can maximize the chances of meeting our operational targets. The management of the risk process also considers the opportunities that are associated with any uncertainty, which allows us to be aware of new prospects that we can capitalize upon. Fundamentally, when we talk about management of risk we need to have a good understanding of the relevance of risk and assessment of priorities. This means that we need to put risks into perspective and focus our efforts on the important ones. Once this is clear, we need to ensure we have a rigorous approach for monitoring and controlling our risks.

Thinking in terms of risk allows us to prioritize our efforts, focusing on the most critical risks and making the best use of the limited resources we have in our facility.

Imagine a process facility organization that has a risk-based approach in their day-to-day work: an organization that at its core manages risk daily, in terms of identifying risk, assessing risk, addressing risk. What would that look like?

Facility Integrity Management
http://dx.doi.org/10.1016/B978-0-12-801764-7.00003-6

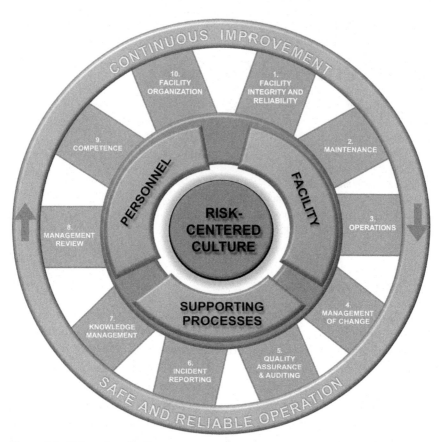

Figure 3.1 *Risk-centered culture.*

We would see a step change reduction in process safety incidents and environmental impacts. The organization would focus on driving straight to the root cause to solve problems in a proactive manner, as opposed to a reactive mindset whereby problems are addressed superficially without spending time to resolve their root causes. The organization would set a target for "zero failures." All equipment and system failures would be addressed with the appropriate amount of effort in accordance with a criticality assessment. Critical failures would be analyzed deep down at their roots in order to find the true defects and eliminate them altogether.

The organization would focus on the important issues that add value and not waste time on the unimportant ones. This is the very nature of risk-centered culture (Figure 3.1). The ownership of the facility integrity initiatives is at all levels in the organization, from management to the shop floor teams and endorsed by senior management.

3.2 RISK

Being able to understand and manage risk effectively is one of the fundamental building blocks of a facility integrity management program. The goal is to achieve a balance between taking risks and the benefits gained from a specific activity. The process of understanding risk includes an appreciation of the specific causes of risk and the associated potential impacts. This may be in the form of safety, health, the environment, a business impact, damage to equipment or a combination. An understanding of risk should also include a sound knowledge base of the process for how we intend to assess risk.

Risk is a subjective concept and is concerned with getting a handle on the uncertainty and limitations of current knowledge and attempting to predict a reliable outcome of an event. Risk cannot be measured physically; it is rather an estimation of the probability of an event occurring along with an appreciation of its consequences.

In mathematical terms risk is defined as the product of likelihood or probability of occurrence of an undesired event and the consequences resulting from its occurrence.

$$\text{Risk} = \text{Likelihood of occurrence of an}$$
$$\text{undesired event} \times \text{consequences resulting}$$

Since risk is a subjective concept, how can we apply it and get the best out of its application for our integrity organization?

The first thing we can do is take this concept a step further by employing a simple matrix of likelihood (or probability) and consequence. This method helps us to quantify risk. Once we can quantify risk, we can appreciate which elements are the most important, which give us the highest return on our investment, which are the most important safety and environmental issues, and which are the most important areas for organizational development. In short, we can prioritize our limited resources, be it manpower or budget, for the high-risk items. The illustration in Figure 3.2 is a simple 5 × 5 risk matrix which is commonly used in the petrochemical and oil and gas industries. The matrix shows that, as the consequences and likelihood increase, so does the risk profile. We can develop a word model for our integrity organization to represent various aspects of risk, including cost, environmental, safety, operational, reputation, etc. and assign a severity rating for each of the five squares for both likelihood and consequence.

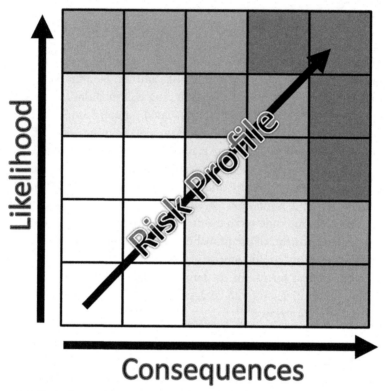

Figure 3.2 *Typical 5 × 5 risk matrix.*

The matrix is shaded from light to dark. The darker areas show the highest risks, which reside in the upper righthand corner of the matrix. These are the areas that we need to focus our efforts on in the first place. We will take this concept further in Chapter 4, where we look at criticality analysis.

3.2.1 Risk Analysis

Risk analysis, as the terms suggest, involves an investigation into the likelihood of undesired events and the consequential damage being caused, with a view to eliminating or minimizing unwanted consequences. It involves identifying initiating events and how they subsequently propagate to result in certain outcomes. Initiating events are concerned with the probability of an event occurring due to failure of systems or equipment.

Risk analysis is concerned with robustness of design and reliability of performance. In this regard it can be viewed as having two main objectives:
1. Safe and reliable equipment design;
2. Development of safeguards to eliminate or minimize loss (of containment) and improve equipment performance.

Table 3.1 Typical probability or likelihood scale for a qualitative risk analysis

Rating	Probability	Description
1	Very Low	Highly unlikely to occur
2	Low	Most likely will not occur
3	Moderate	Possible to occur
4	High	Likely to occur
5	Very High	Highly likely to occur

Table 3.2 Typical consequence scale for a qualitative risk analysis

Rating	Consequence	Cost
1	Very Low	No increase in budget
2	Low	<10% increase in budget
3	Moderate	10–20% increase in budget
4	High	20–30% increase in budget
5	Very High	>30% increase in budget

A risk analysis can either be quantitative or qualitative or a suitable combination of both, which depends on the aims and objectives. The central difference between a quantitative and a qualitative risk analysis is that a quantitative risk analysis establishes a numerical scale, such as a probability of a risk occurring. Qualitative analysis, on the other hand, uses a descriptive scale to measure the probability of a risk occurring, such as Low, Medium and High. Tables 3.1 and 3.2 respectively show typical scales for probability and consequences for a qualitative risk analysis. The consequence scale may be tailored to meet specific criteria when assessing risk. As an example, consequence with respect to cost has been presented in Table 3.2.

A quantitative risk analysis may be required to analyze technical risks such as cost effectiveness or fitness for purpose of equipment designs, which may be required for making decisions to accept a known risk or mitigate it. Qualitative risk analysis may be required when dealing with the softer organizational-based risks or higher level risk analysis.

Once a risk analysis is completed and high-level risks have been identified, it is necessary to instill control measures so that risks can be managed. Control measures should deal with the specific issues and the effort expended should be proportionate to the concerned risk. We can instill control measures in a number of different ways. This may be in the form of an engineering solution to design the risk out or alternatively through the development and implementation of a procedural control, or a combination of both.

Our priority when implementing control measures is first and foremost to prevent the risk; if this is not possible we can reduce the risk to an acceptable level at its source. Our efforts should focus on the development and implementation of an engineering solution to eliminate the risk altogether, then to move down the hierarchical risk ladder to control measures that are more difficult to enforce and maintain such as facility teams ways of working and facility operational controls. As a last resort, we should consider an emergency response plan.

3.2.2 Risk Management

The British Standard BS4778 defines risk management as "the process whereby decisions are made to accept a known or assessed risk and/or the implementation of actions to reduce the consequences or probability of occurrence" [3.1]. The underlying principles in risk management therefore are to identify and manage risk to an acceptable level. This helps us to prioritize our resources, dealing with the most important or highest risks first with the right level of effort.

The process of risk management essentially involves the following activities:

- Identification of risks;
- Analysis or assessment of risk;
- Elimination or reduction of risk;
- Development and implementation of a management strategy for control and mitigation of the risk;
- Monitoring and auditing the management strategy for improvement.

Effective management of risk starts from an understanding of the initiating events, their root causes and potential consequences. Management of risk is a central concept for our risk-centered culture within our facility. In this regard our aim is threefold:

- To identify and analyze all risks;
- To lower risk severity by driving risks down the risk profile (Figure 3.1);
- To prioritize our limited resources on the most important things.

Now that we understand one of the fundamental principles behind risk-centered culture, we are ready to focus on the application of the risk concept in order to add value. Our facility integrity efforts are owned and proactively driven by the organization as a whole from the shop floor teams to senior management. We must assess all risks by balancing the appropriate amount of effort so that we are effective and efficient in addressing our integrity problems, developing and implementing improvements while managing cost.

We must consider all potential risks that may at first glance appear unimportant. Until we make an assessment of risk we may not understand its significance. In order to illustrate this point we shall revisit a case study (from section 2.6.1, Case Study – ConocoPhillips, Humber Refinery) to help understand this concept and appreciate the importance of each individual in the organization thinking in this regard. It is evident that by overlooking risk events can unfold and result in a catastrophic failure.

3.3 CASE STUDY – THE SILO COMPANY

The design and installation of the new water injection point on the saturated gas plant (SGP) on the Humber Refinery was not subject to a robust management of change assessment. There was no or little consideration that a simple change to the gas piping would and could have had such a major impact. If such an assessment had been carried out, it would have been evident that there was a significant corrosion risk that the injection point introduced for the downstream pipework [3.2].

Similarly, no or little risk assessment was made regarding the changes in the use of the water injection point from continuous to intermittent over the lifetime of the facility. This is an operational requirement which was not made clear to the facility integrity team. The frequency of use of this injection point was a major factor in the rate of the corrosion of the pipework. The assessment would have uncovered that, during periods of continued use, the corrosion rate would be significantly increased. The assessment would have pointed to the requirement for more frequent inspections of the pipework in order to adequately monitor the integrity of the pipe.

The lack of focus on identification and management of risks was clearly a critical factor. Risks that would likely have been perceived as significant by some of the facility groups were not shared with the concerned groups. Equally at fault were significant failings with regard to the lack of communication, which defeated barriers along the way to the incident. The changes to the frequency of use of the new water injection system were not communicated outside of the facility operations team. As a result, there was a belief within the other facility teams that the new water injection system was only in occasional use and did not constitute a corrosion risk. During a subsequent detailed inspection of specific human factors, it was found that communications were largely "top down" instructions, rather than seeking to involve the workforce.

There are a number of key learning points that can be brought out of this case study. First, it is clear that one of the key contributing factors that led to the incident was that the facility departmental teams did not communicate sufficiently and effectively with each other. Effective communication is an important part of any facility integrity program. The accurate recording and effective flow of information concerning facility equipment and systems is essential for accident prevention and facility performance. Such communication should actively involve the workforce in the prevention of major accidents as part of an adequately resourced process safety management system.

Silo mentality fosters an environment where groups or individuals do not share information or knowledge with others openly or freely. The result is a self-destructive environment which is inefficient and ultimately contributes towards a failing facility culture. We need to make sure that information flows without restriction between our facility groups to ensure that our integrity organization is effective.

If we underestimate the importance of proper risk management and open communication between facility groups, we leave ourselves exposed to the potentially catastrophic effects of critical equipment failure at a facility. The events described in the case study could have easily been prevented. We need to work within an integrated team and create an environment that encourages unrestricted and proactive communication, in addition to fostering an environment that encourages proactive identification and mitigation of risk within each of the facility groups. These are essential ingredients for a risk-centered culture.

3.4 MOVING TOWARDS AN INTEGRATED APPROACH

In creating a risk-centered culture we need to move away from a silo-based way of thinking and silo approach to risk management. This means that, instead of the operations team focusing only on operational risks and the maintenance team on maintenance risks, an approach where risks are identified and reviewed jointly in an integrated team is adopted. In this way we can avoid addressing risks in silos, which is hugely restrictive and promotes an environment for inefficiency and failure.

Our objective as a risk-centered organization is to bring together the three core integrity elements within the Facility Integrity Excellence Model: the facility integrity and reliability group, the maintenance group and the operations group, as we can see in Figure 3.3. There are huge synergies to be had by overlapping the skillsets and experiences of the three core integrity groups.

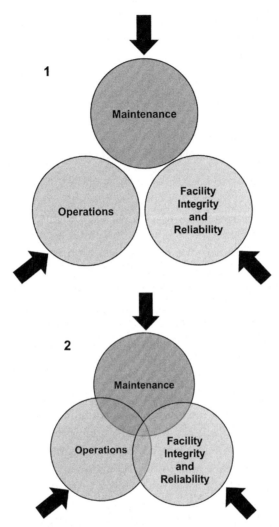

Figure 3.3 *Moving towards an integrated approach.*

Moving towards an integrated approach requires our integrity organization to take a step back and take a critical look at ourselves in the mirror. We need to understand how we are behaving today as an integrity organization and what impact that is having on our performance. Do we have the traits of a silo organization as in Figure 3.1, scenario 1? Or are we somewhere between scenario 1 and 2?

The organizational culture is the foundation for communication between the different integrity groups at the facility. It is imperative to ensure that the right culture is in place to build the integrity organization upon.

Cultural improvement will require meaningful changes to take place in the current ways of working within the organization.

Addressing the organization's culture is a major undertaking; however, a step change in the way the organization thinks and reacts when it comes to integrity is imperative in order to ensure the FIEM succeeds in the long term.

3.5 MORE ON RISK-CENTERED CULTURE

Risk-centered culture (RCC) is the term given to the heart of the Facility Integrity Excellence Model, which represents the organizational culture. Culture is a key component of the concept of risk-centered culture, which considers the way the facility organization's employees behave. The integrity organization must have the right beliefs, mindset and way of working in order to be successful. When the major oil and gas incidents of the 21st century are analyzed, such as Piper Alpha, it is clear that poor discipline has led to inadequate risk management, resulting in a major incident that could have easily been prevented.

Risk-centered culture is an ideal that describes the values, beliefs, knowledge, attitudes and understanding about risk shared across the organization and driving us towards a common purpose. RCC aims to achieve a "zero failure" culture that safeguards the site facility, employees and community and maximizes returns on the facility. Risk-centered culture is the center of the Facility Integrity Excellence Model. It is the lifeblood of the organization that ensures the core integrity concepts are ingrained into the organization from the shop floor level to the top management.

We have discussed risk and management of risk at length. This is an integral part of risk-centered culture. However, we need to consider that a key part of our efforts to identify and address risk is to be as efficient as possible with deployment and utilization of our resources. Facility resources, our people and materials are very precious and usually in short supply and we need to make sure we are getting the best value out of them. We can do this by focusing on what is important, as we have shown in Figure 3.2. Imagine how much more effective we can be in our facility operation if we had each and every team adding optimum value, prioritizing their efforts to deal with the critical items and eliminating wasted time.

As we move into the following chapters, we can see a number of techniques and tools that we can employ to optimize our resources, such as risk-based methods.

Now that our facility organization is focusing their efforts on what's important, as a risk-centered culture we also need to ensure the organization is equipped with the necessary skills and tools to be able to perform duties satisfactorily. We need to be a learning organization. (We will explore this concept in detail in Chapter 8, The Facility Integrity Organization.)

There are a number of key areas that we can focus on to bring about a risk-centered culture:

- Embed risk in the core integrity processes and systems, such as planning meetings for operations and maintenance, performance assessment of facility equipment, root cause analysis for failed or failing equipment, etc. Implement gated reviews or other independent risk reviews in order to embed risk-centered structure into the organization.
- Conduct risk management education and training programs for employees and managers and customize them based on individual roles within the facility organization. Embed risk awareness into induction programs.
- Strengthen formal communication lines and encourage informal communication between the core facility integrity groups.
- Use a common risk language. Create an environment where the organization is comfortable with talking openly and honestly about risk. This can be promoted by developing and educating the various facility teams to use a common risk vocabulary, which will help to encourage shared understanding.
- Conduct a critical review of the existing facility integrity flow of information and assess how effective the feedback loops are within the facility. In doing so, identify and close the gaps.
- Promote facility integrity requirements to the internal and external stakeholders through a publicity campaign, weekly newsletters, facility intranet, knowledge sharing sessions, "lunch and learn" initiative, etc.

3.6 CHANGING AN ORGANIZATION'S CULTURE

"A change sticks when it seeps into the bloodstream of the corporate body" [3.3]. Until new behaviors are ingrained into the shared values of the organization, they are subject to deterioration as soon as the pressure for the change is lifted. It is the culture of the organization that is the target for change in order for the implementation of the FIEM to be effective. We shall explore in detail how to go about implementation of major change, such as a new integrity management system, in Chapter 10.

In order to change the culture, the participation of the whole organization is required. The process of identifying, analyzing, evaluating and managing risks should not be solely left in the hands of senior personnel. Although senior management has the overall say in deciding the destiny of the business, the managers might not possess all the knowledge about the risks facing the business. In order to embrace cultural change, it is necessary to create an environment that recognizes and rewards people for paying attention to risk. This includes having courage and knowing how to challenge the current ways of working constructively.

Some important factors of changing an organization's risk culture are:

- Creating a culture of constructive challenge.
- Embedding risk performance metrics into motivational systems.
- Establishing risk management considerations in talent management processes.
- Positioning individuals with the desired risk orientation in roles where effective risk management is critical.
- Reinforcing behavioral, ethical and compliance standards.
- Integrating risk management lessons learned into communications, education and training.
- Holding people accountable for their actions.
- Refining risk performance metrics to reflect changes in business strategy, risk appetite and tolerance.
- Redeploying individuals to reflect changes to business strategy and priorities.

Risk-centered culture within the FIEM requires integrity processes to be embedded at all levels of the organization, top- down and bottom- up, from the policy to the strategy and finally to the work procedures and standards that operate within the company.

As we start to develop a risk-centered culture, we will find that our behaviors will move from a silo mentality towards proactive cooperative relationships between operations, maintenance and engineering, which is of paramount importance. We will see that input and ideas are offered from the different facility groups with a "don't just fix it, solve it" mindset.

This requires a culture that can permit honest mistakes, but is disciplined in the event of negligence, which requires trust and honesty. There is a willingness to keep an open mind, and rather than placing blame, a mentality to find solutions.

CHAPTER 4

Facility Integrity and Reliability

Contents

4.1 INTRODUCTION

An independent department that manages facility integrity and reliability (FI&R) as a stand-alone function in an oil and gas or petrochemical facility may be an alien concept to some. However, as part of a successful facility integrity management system and the Facility Integrity Excellence Model (FIEM), it is an essential spoke in the wheel (Figure 4.1). FI&R has a critical role to play in order to safely operate the facility and to maintain a desired product throughput, product quality and acceptable equipment availability.

An FI&R function in a facility can be approached in a number of different ways from organization to organization, but essentially the basic ingredients remain the same.

The *reliability* element of FI&R is mostly concerned with rotating equipment such as pumps, compressors and turbines, for example. It focuses

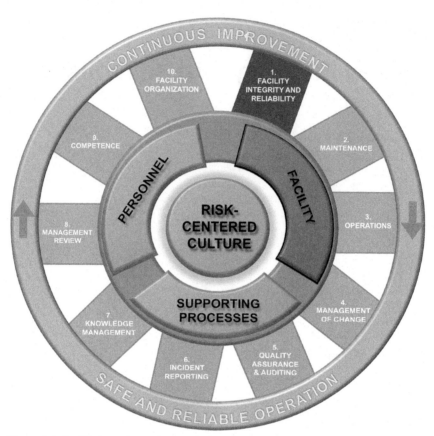

Figure 4.1 *Facility integrity and reliability is a crucial element.*

on assessment of the performance and the optimization of the equipment in order to ensure availability meets the company's business objectives. We shall discuss reliability in detail in section 4.6.

The *facility integrity* element of FI&R is mostly concerned with static equipment such as piping, heat exchangers and vessels, for example. Static equipment is concerned mainly with ensuring that there is no loss of containment. It deals with a magnitude of material degradation mechanisms in their operating environments such as corrosion, erosion and creep. It focuses on the appropriate application of various inspection technologies and remedial measures that can be taken to prevent such degradation mechanisms. Careful analysis of the inspection data that is collected helps in deciding on the appropriate remedial measure, which may include redesign, chemical inhibitors, tailored maintenance programs and replacement strategies. Inspection is an important feature of FI&R and the FIEM and is detailed in section 4.7.

4.2 THE FACILITY INTEGRITY AND RELIABILITY FUNCTION

The FI&R department or function aims to maximize facility uptime for static and rotating equipment. To break the function down into its two constituent parts, the reliability function is concerned with predicting and avoiding failures and the integrity function involves close monitoring of known static degradation mechanisms and taking proactive action, in many cases through corrosion management activities.

Facility integrity and reliability strive to ensure the facility is operated safely and facility systems and equipment performance is optimized in order to deliver production targets. Essentially facility integrity and reliability are all about maximizing profits while maintaining acceptable standards of safety and performance.

A typical organization chart showing what a FI&R function may look like on an oil and gas or petrochemical facility is shown in Figure 4.2. Essentially the function is divided into two groups, reliability and integrity. We mentioned earlier that reliability is mainly concerned with rotating equipment or equipment with moving parts, whereas integrity mainly deals with static equipment.

The reliability group generally comprises reliability engineers, a root cause analysis (RCA) team (although during RCA investigations the integrity team, among others, must be consulted), and a condition monitoring (CM) team. The reliability engineering competence provides expertise, in general through the use of *reliability centered maintenance*, which we shall discuss in Chapter 5, Maintenance Management, to focus the condition monitoring efforts on the critical equipment. The condition monitoring team may use a number of technologies and tools in order to ascertain equipment performance, which we will also review in detail in Chapter 5.

The integrity group is generally broken down further into two groups, the inspection group and the corrosion or material degradation group. The inspection group is made up of a number of facility inspectors that are tasked with inspection of the facility, focusing on static equipment for signs of equipment degradation using a number of different technologies and tools. The corrosion or material degradation group provides expertise to the inspectors as to where their effort needs to be prioritized. This is achieved by a variety of methods that include mathematical models and simulators that are able to predict where corrosion occurs on the facility and to what extent.

Facility integrity and reliability covers a wide range of concepts, all of which are essential to ensure the facility equipment and systems perform their duty as required. A fundamental element of any FI&R function is the management of facility equipment and systems data. This is accomplished

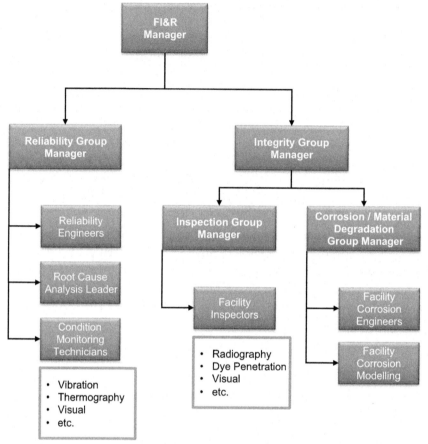

Figure 4.2 *The facility integrity and reliability department.*

through databases and computer software packages or even hard copies in some cases. The data is collected from a wide range of sources, which include condition monitoring and inspection data, which may be live data or historic trends. It is essential that this data is from the right source, that it is reliable and up to date, stored effectively and shared with the relevant parties. We will discuss management of knowledge in Chapter 7.

4.2.1 Failure of Equipment and Systems

Essentially, FI&R is concerned with avoiding failures of equipment and processes by proper design and careful operation of the equipment by trained personnel. Close monitoring of the facility and its equipment offers the

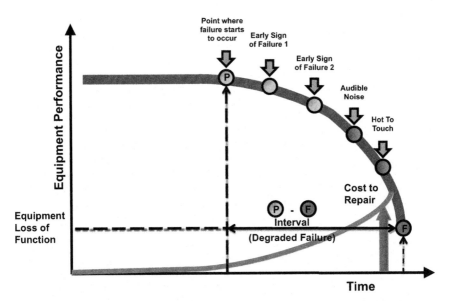

Figure 4.3 *P-F curve – equipment failure.*

opportunity to avoid failure mechanisms and to detect early signs of equipment deterioration that can be remedied before a failure occurs. A failure-free environment allows the facility to be utilized more effectively and production output to increase, ultimately producing a higher return on sales.

Figure 4.3 shows a graphical representation of the behavior of equipment failure leading to the ultimate loss of function – the P-F curve [4.1]. It is important to understand how equipment fails and how early detection of a failure provides time to plan and schedule maintenance to resolve the failure before the failure becomes catastrophic, in order to minimize interruption to production.

As an incipient equipment failure starts to manifest, it deteriorates to the point where it can be detected. This is known as the point of onset of failure (P). This may be an incipient failure in its initial stages as it may not be detected. As the failure progresses, the incipient failure may develop into a degraded or critical failure as the performance of the equipment progressively worsens and the equipment is unable to deliver its original design intent.

Degraded failures may be detected by using a number of failure-finding techniques such as inspection, condition monitoring and process diagnostics. By early detection and resolution of equipment failure, there can be major savings made by eliminating the potential for failures to occur that can lead to major incidents at the facility.

As the equipment approaches the total loss of function (F) point in the curve, the resolution upon detection of the failure is usually a reactive repair. Ultimately the failure progresses to a point where there is total loss of function (F).

The duration between the point of onset of failure (P) and the functional failure (F) is commonly known as the P-F interval [4.1]. This is the window of opportunity to carry out monitoring and inspection in an attempt to detect the failure and address it. Detection and resolving equipment failures before they become critical can add a great deal of value to the facility.

The vast majority of failures that occur are random, which means that they may occur at any time. We shall now look at the dominant failure patterns which were researched and presented by Nowlan and Heap during the 1970s [4.2].

4.2.2 The Bathtub Curve

Until relatively recently, it was believed that all equipment would exhibit wear-out characteristics of a typical "bathtub" curve, which is a popular concept in reliability literature. This type of curve has three identifiable regions: an infant mortality region, where there is a period immediately after manufacture or overhaul with a high probability of failure; a region with a constantly low failure probability rate; and a wear-out region in which the probability increases quickly with the age of equipment.

During the 1970s and 1980s extensive research was conducted within the aviation industry on failure characteristics by F. Stanley Nowlan and Howard F. Heap, both senior executives from United Airlines [4.2].

The research turned up some interesting results and concluded that 89% of the items that were analyzed had no wear-out zone. This means that, for these items, after a certain time the probability of failure continued at a constant rate. In other words, their performance could not be improved by intervention. It was found that the conditional probability curves fell into six basic patterns. No matter the equipment type, its reliability pattern would match one of the six basic patterns. These reliability patterns are illustrated in Figure 4.4 [4.2].

Bathtub curves or failure probability patterns account for changes in failure rates over the life cycle of the equipment. The research also shows us that the probability of a failure can be influenced. If we are aware of the pattern of failure, we can tailor our FI&R and maintenance efforts to select the most appropriate maintenance strategy in order to drive down the failure rate to acceptable levels.

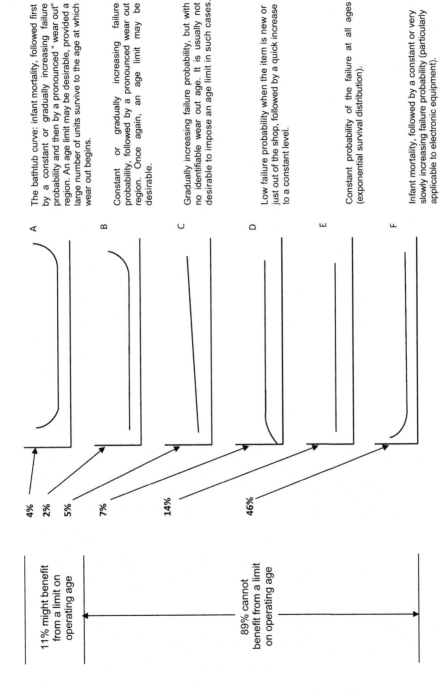

A — The bathtub curve: infant mortality, followed first by a constant or gradually increasing failure probability and then by a pronounced "wear out" region. An age limit may be desirable, provided a large number of units survive to the age at which wear out begins.

B — Constant or gradually increasing failure probability, followed by a pronounced wear out region. Once again, an age limit may be desirable.

C — Gradually increasing failure probability, but with no identifiable wear out age. It is usually not desirable to impose an age limit in such cases.

D — Low failure probability when the item is new or just out of the shop, followed by a quick increase to a constant level.

E — Constant probability of the failure at all ages (exponential survival distribution).

F — Infant mortality, followed by a constant or very slowly increasing failure probability (particularly applicable to electronic equipment).

4%
2%
5%
7%
14%
46%

11% might benefit from a limit on operating age

89% cannot benefit from a limit on operating age

Figure 4.4 The bathtub curve.

4.3 REACTIVE APPROACH TO FACILITY INTEGRITY MANAGEMENT

Historically, the way the oil, gas and petrochemical industry has tended to respond to equipment failure is with a "fix it when it breaks" mentality. In other words, it has been reactive in its approach. This is in contrast to working to detect equipment failure and working to prevent it – a proactive approach.

In general there was little effort directed towards the detection of early signs of equipment failure and action only took place once the failure actually started to threaten production output. Efforts in failure analysis tended to be at a high level, usually focused around human error and mechanical failure. Meanwhile, more indirect causes of failure such as deviation from original equipment manufacturer's (OEM) specifications and improper operation of equipment may not have been considered. We can see in Figure 4.5 that the reactive approach to maintenance management can be punishing cycle.

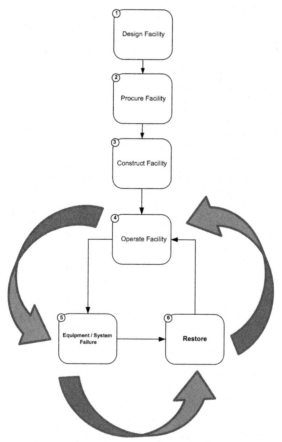

Figure 4.5 *The reactive approach to maintenance management.*

Following the installation of a new item of equipment on a facility, the equipment enters into an operating cycle, whereby it operates for a period of time, fails and corrective maintenance is administered to repair or replace the failure. Once the equipment is reinstated, the operation continues and ultimately this cycle continues. There is often no or at least a minimum amount of work to predict degraded failures before they become catastrophic and reach point (P) in the P-F curve (Figure 4.3).

The nature of this paradigm means that maintenance costs are unnecessarily high and reliability performance is low.

4.4 MOVING FROM A REACTIVE TO A PROACTIVE MINDSET

Today the industry landscape has changed significantly. Competition is at its peak and it is more important than ever that the facility be online as required in order to deliver maximum production. Our approach to integrity management must center around being proactive rather than reactive. Each and every hour of production must be accounted for and our facility must be reliable in order to remain competitive. The workflow process in Figure 4.6 shows us the additional steps as we start to move from a reactive to a proactive approach.

If facility equipment fails we spend time to carry out root cause analysis to understand the failure modes so that we can eliminate or mitigate them. We are starting to take control of our facility equipment with fewer failures now that we are addressing the root cause.

Accurate and comprehensive data is the foundation of any improvement effort; therefore the next step is to take precautionary measures before a functional failure by monitoring and capturing equipment and performance data from the operating facility and from degraded failures. This step effectively allows us to manage our failures much more effectively and improve facility uptime. We are able to find the defects in equipment and systems at their root and ultimately address these in the design, procure and construct stages. A reliable plant and well-organized maintenance function will result in lower overall maintenance costs.

4.5 RISK-BASED METHODS

We have already discussed the fact that traditional maintenance and inspection activities have tended to be time based. Many oil, gas and petrochemical companies are moving to risk-based methods (RBMs).

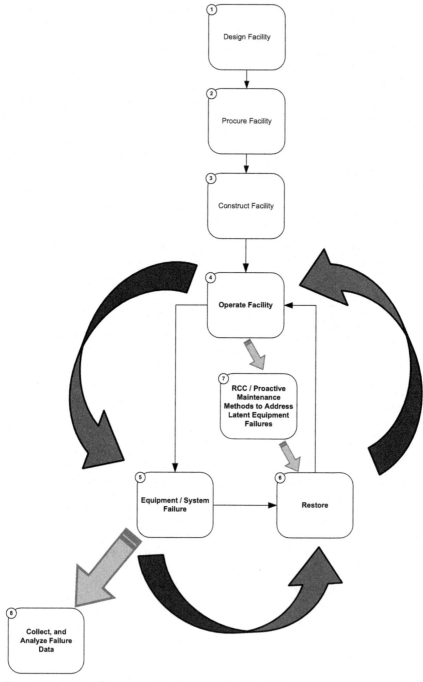

Figure 4.6 *Moving from a reactive to a proactive mindset.*

Adopting a risk-based approach is a key feature of the FIEM. RBMs provide a structured risk management program to minimize facility downtime, and maximize safety performance. RBMs also allow us to identify and take advantage of opportunities to improve the facility performance.

There are a number of important advantages in applying RBMs, including:

- Improvements to safety and reliability of facility equipment;
- More robust and reliable process for maintenance and inspection tasks;
- Improvements to effective utilization of resources;
- More effective prioritization of maintenance and inspection tasks;
- Improvements to quality assurance and auditing.

There tend to be three fundamental risk-based methodologies. These methodologies are intended for specific types of equipment and systems, as follows:

- *Risk-based inspection* (RBI), an inspection-based methodology that examines static facility equipment for material degradation mechanisms and assesses the probability and consequence of failure. RBI then ranks the risks so that optimized inspection plans can be developed in order to mitigate the risks. RBI focuses on static facility equipment such as piping and pipelines, vessels, heat exchangers, tanks, boilers. RBI is presented in section 4.7.
- *Reliability centered maintenance* (RCM), a risk-based method that focuses on ensuring that facility equipment and systems are functioning as per their design intent throughout their design life. RCM is also employed to further optimize and improve equipment performance at the facility as well as the associated optimized maintenance cost. RCM is defined by the technical standard SAE JA1011, Evaluation Criteria for RCM Processes [4.3]. RCM generally applies to rotating and reciprocating equipment. It is also applied to heating and ventilation, and electrical equipment. RCM is discussed in detail in Chapter 5.
- *Instrumented protective function* (IPF), a risk-based approach that focuses on safeguarding the integrity of facility instrumentation and control systems. The IPF method is associated with safety instrumented systems (SIS), which comprise an engineered set of hardware and software controls used on critical process control systems. IPF tends to be applied to control systems such as fire and gas detection, emergency shutdown and facility process control systems, etc.

4.6 RELIABILITY

Reliability can be described in a number of ways:
- Equipment that is fit for a purpose for a specified duration of time under specified conditions;
- The probability of an item of equipment to perform a required function under stated conditions for a specified period of time;
- The ability of an item of equipment to consistently perform its intended function, on demand and without degradation or failure;
- Failure-free performance within a specified timeframe, under a specified environmental;
- Mean time between failure;
- Quality over time.

Ultimately, reliability is about preventing equipment failure by careful operation of the equipment and proper design in a specified environment and over a specified timescale. It is driven by cost since preventing failures costs money and improvement decisions are always about money.

Reliability is not a physical characteristic; it cannot be felt, touched or smelled. It is, however, real and of critical importance in order to ensure the facility operates effectively. Reliability can be measured but it takes time for accurate reliability measurements; the longer the measurement, the more accurate the reliability measurements.

Reliability can be defined as the probability that an item of equipment can perform its intended function under specified conditions for a specified time span. The time span is often referred to as mission time.

$$\text{Reliability} = e^{-t/\text{MTBF}}$$

where t = mission time and MTBF = mean time between failure. The exponential failure distribution describes chance failures and it is used to simplify the multiple mixed failure modes for system failures.

MTBF is an important term because it quantifies reliability performance and can be informative as to how reliably the facility equipment is performing.

$$\text{MTBF} = \frac{\text{Total functioning life of all equipment on a Facility}}{\text{Total number of failures within the Facility during the measured interval}}$$

MTBF is used to measure reliability performance for repairable systems since it focuses on the time span between failures, which can be used to plan preventative maintenance. If we have a short MTBF in relation to the mission time, we can say that the equipment in question is unreliable. If the MTBF is long in relation to the mission time, we can describe the equipment as reliable.

Another important quantification of reliability is mean time to repair (MTTR). It is a basic measure of the maintainability of repairable equipment items and measures the average time that is needed to repair a failed item of equipment or component:

$$\text{MTTR} = \frac{\text{The total corrective maintenance time}}{\text{Total number of corrective maintenance actions}}$$
$$\text{(during the measured interval)}$$

We can now represent availability as follows:

$$\text{Availability} = \frac{\text{MTBF}}{\text{MTBF} + \text{MTTR} + \text{PM}}$$

where PM is the lost time for preventive maintenance work.

If we focus on increasing MTBF or decreasing MTTR, we will see an improvement in availability.

4.6.1 Reliability Engineering

Reliability engineering focuses on costs of failure caused by system downtime, which includes cost of spare parts, equipment repair, equipment overhaul, personnel and equipment warranty.

The goal of reliability engineering is to carry out an assessment as to the reliability of facility equipment and identify potential areas for improvement. This includes improvement not only in equipment design but also in terms of how it is operated and how it is maintained. From a pragmatic perspective, all failures cannot be eliminated; therefore reliability engineering also focuses on identification and mitigation of the high-probability failures and their effects.

Reliability engineering is a continuous process which starts at the conceptual stage of the facility design and continues throughout all stages of the facility life cycle. It involves a number of different tools and technologies in the process. The goal should be to address potential reliability problems as early as possible in the facility life cycle so that cost can be minimized. As we can see in Figure 4.7, if we catch reliability problems early in the facility life cycle, at the concept design or detailed engineering design stage,

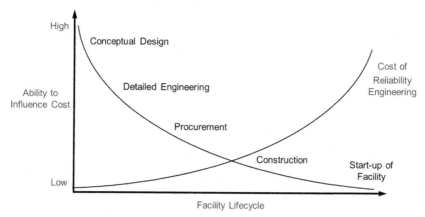

Figure 4.7 *Cost of reliability engineering.*

the associated reliability engineering costs are orders of magnitude less than later on in the life cycle at the construction or operating stages.

Reliability engineers should have a detailed understanding of the cost of unreliability and have a handle on where the focus needs to be to address it.

What is the cost of unreliability of your facility?

How much does it cost for your facility to be offline for 1 hour?

An understanding of unreliability and how to answer these questions is an important step along the journey to the FIEM.

4.7 INSPECTION

The facility inspection group is responsible for a number of critical activities which center around the monitoring and performance assessment of static equipment on the facility. This may be achieved through many activities and inspection technologies.

Inspection activities involve retrieving static equipment information from the facility and making an assessment as to the rate of material degradation so that preventive maintenance can be carried out ahead of failure. Material degradation may be in the form of corrosion, erosion or creep, for example. There are many different degradation mechanisms at play, which all require special attention to mitigate. The inspection group is tasked with identification, assessment, prediction and the development and implementation of mitigation measures in order to deal with these degradation mechanisms.

Depending on the severity of the risk, there are many methods, tools and technologies at the disposal of the inspection group to address them. Manual inspection techniques include visual inspection to inspect surface condition of equipment and to revisit defects that have previously been identified to check for further deterioration. The visual inspection data is compared against a standard benchmark, so there is a basis for comparison. The frequency of carrying out visual inspection is dependent on a number of parameters. These can include an understanding of how severe the service is: for example, if the service is highly corrosive a much more frequent inspection is required than for a noncorrosive service. Statutory examinations as imposed by regulatory bodies also stipulate inspection frequencies, which need to be adhered to.

Modeling of facility equipment corrosion trends is a key factor in the development of an inspection program. Corrosion trends take the existing facility design and service data and provide a computational model that can predict corrosion. This data is used to develop inspection requirements.

The inspection group comprises a team of inspectors that are tasked with inspection and monitoring the facility, as defined by an inspection program, on a daily basis. The data is reviewed and assessment made as to where the key focus areas are.

The inspection group may also include corrosion engineers as part of the FI&R group. Material degradation mechanisms, such as corrosion, are of major concern to a facility. It is imperative that the FI&R group have a clear understanding of material degradation on the facility and be able to manage the risk appropriately. Corrosion engineers are able to predict focus areas for corrosion and corresponding rates of corrosion on the facility. Computerized database systems are able to simplify corrosion data management by modeling corrosion trends and predicting high risk areas.

Data is retrieved from the facility by means of non-destructive testing (NDT) techniques at established test points as derived from an inspection program. These data are often linked to an isometric model of the facility. Static equipment wall thickness measurements are taken by the inspectors to verify model predictions against established test point locations. There are a number of commercially available software titles to achieve this goal.

The inspection team may opt for onstream monitoring on the facility. This option is a more expensive alternative, which is able to produce live results and is particularly employed for more critical areas. Onstream monitoring is useful since it enables a live measurement of the rate and extent of corrosion or material degradation which has occurred while the facility is still in service.

The inspection method chosen must be capable of providing an accurate and reliable indication of corrosion or material degradation rates and the current condition of the plant. A number of technologies may be used that have been tried and tested in the industry for many years, including:

- Ultrasonic examination, a nondestructive technique that utilizes portable ultrasonic flaw detectors or digital/analog thickness meters to measure wall thickness of static equipment such as piping.
- Electrical resistance corrosion probes, which can measure corrosion rate during operation without probe removal. The operating principle of the probe is based on the change in resistance of the probe element as it is exposed to corrosive conditions. Corrosion probes are inserted into a facility process unit, usually through a valve, which is exposed to the process service fluid. The probe condition can be monitored using either an automatic monitoring system, by means of portable instruments or by removing it from the process unit for physical examination.
- Digital radiographs, a form of X-ray imaging which can provide an immediate image preview. Digital radiographs can be used to determine the wall thickness of static equipment.

Early detection of static equipment failure is of paramount importance to facility inspectors. It provides time to plan and schedule maintenance to resolve the failure before the failure becomes critical. In order to achieve this goal it is important to prioritize efforts given limited resources on the facility. Risk-based inspection is an industry norm that strives to achieve this goal.

4.7.1 Management of Integrity Data

In order to facilitate effective management of inspection and corrosion data, it is often necessary to employ a computerized inspection and corrosion management system (ICMS). An ICMS draws together all information relating to inspection and corrosion on the facility. This allows the FI&R group and the facility management team to make more informed, data-driven decisions relating to the integrity of the facility. An ICMS plays a key role in the FIEM, in that it improves communications between the different functions on the facility including operations, FI&R and maintenance.

An ICMS also enables streamlining of workflow processes and provides an auditable documented trail. This includes inspection records and historic integrity performance data such as corrosion degradation, which can provide a solid basis for predicting equipment degradation across the facility.

It is imperative to understand facility equipment corrosion behavior for critical equipment. An ICMS is a cornerstone to enable effective modeling of equipment degradation mechanisms at play. It is essential that the data uploaded into an ICMS is accurate and consistent in order to produce credible and reliable predictions.

A computerized inspection and corrosion management system is a database that is efficient and effective in the storage and retrieval of integrity data. It should be simple to use and flexible to capture changes to facility processes or equipment and provide a robust reporting function. It should be capable of interfacing with other inspection and corrosion software utilized on the facility. This is in order to eliminate any double handling of inspection and corrosion data and revision management, which can be a source of error. Furthermore it should be able to interface with the maintenance management system (MMS). This is because there are also a lot of equipment history data stored within an MMS. Collectively the MMS and ICMS can provide a comprehensive picture of facility equipment performance based on historic data.

Finally an ICMS must be designed in order to support a risk-based inspection program (RBI). RBI is a critical process for FIEM since it provides a firm foundation for managing the integrity of static equipment and it also supports and drives the fundamentals of a risk centered culture.

4.7.2 Risk-Based Inspection

Risk-based inspection is an approach that aims to minimize equipment downtime and prolong equipment life on a facility. An effective RBI program can substantially increase the effectiveness of an existing facility integrity program. It reduces risk associated with equipment failure, which in turn reduces risk to health, safety and the environment (HSE). Since it is based on management of risk, it can also be effective in prioritizing facility resources, which is a key component of a risk-centered culture and the FIEM. RBI focuses on static equipment which includes piping, pipelines, vessels, drums, and heat exchangers, among many others.

The main benefits of a risk-based inspection program are:
- Confidence in facility equipment performance;
- Effective visibility and management of equipment degradation mechanisms;
- Reduction in facility downtime;
- Overall cost savings over "traditional inspection" methods;
- Adherence to HSE compliance standards for the industry.

The basis of RBI is risk assessment. RBI methodologies tend to differ based on the depth and type of risk assessment that is performed. Risk assessment methodology may be qualitative, quantitative or a combination of both. Quantitative methods tend to yield results that are fully quantified in terms of absolute failure probabilities and consequences. Qualitative methods tend to follow a scoring type assessment from using generic probability values. In general, quantitative methods are more expensive to perform because they require more resources and time to develop.

Essentially the development of an RBI program for a facility or part of a facility involves either an expert review or employment of a fit-for-purpose RBI software package, or a combination of both.

4.7.2.1 RBI Software System

There are numerous software systems available in the petrochemical and oil and gas industry to support the development of an RBI program. RBI software requires facility integrity data to be collected ahead of time and uploaded into the software system. It may be difficult in some cases to collect this data for older facilities with aged equipment and systems.

Data collection and accuracy is a critical factor in RBI. A comprehensive and quality data set is paramount in order to ensure the output of the RBI program is reliable. Data collection is expensive and there must be a balance between costs committed to the RBI program and the return on investment back to the facility. In most cases, data collection can be done by nontechnical staff, which helps to keep the costs down, provided it is carefully managed.

Review and sign-off of the inspection data must be done by qualified and experienced engineers in order to ensure reliable and quality data. RBI engineers are part of the FI&R group and should have a corrosion or materials background, be conversant with the facility process and be knowledgeable about inspection techniques and technology. These individuals play a vital role in the development and also the continued operation of an RBI program. They are required to carry out frequent reviews of the RBI software model, accuracy checks, and analysis of degradation mechanisms at play. The RBI software program is able to compute likelihood assessments, consequence assessments and risk ranking of failure modes in conjunction with the facility data and material and fluid property database. Inspection plans can be produced within the software, which contain an inspection task and a frequency of inspection. The results of the RBI assessment are reviewed by the RBI engineers to ensure a robust output before implementation in the facility.

RBI software systems can fast track the development of corrosion and degradation risk assessments and produce inspection plans very quickly. They are able to perform bulk processing of data, which saves time and resources. The output of an RBI software system is auditable and repeatable, which is an important element of the FIEM.

RBI software systems are also integrated with the ICMS and even in some instances form part of the same system. This is important for continuity of data and revision control. Furthermore, since the development of an RBI program involves the creation of a complete register of facility equipment, this is uploaded to the RBI software systems and is able also to link to the computerized maintenance management system (CMMS), which we discuss in Chapter 5.

4.7.2.2 RBI Expert Review

The second methodology is an "expert review." This requires the assembly of an expert review team. The team must have the correct disciplines, skills and experience required in order to ensure that all the inputs into the review are understood in the interest of producing a quality product. The team is brought together to carry out the RBI assessment which requires a large investment of time. It is important that the RBI assessment is performed in an environment that is distraction free in order to focus on the task. The RBI assessment comprises a detailed review of each item. This includes a thorough review of the design and service history of each item at first, in order to come to reasonable conclusions about failure likelihood and consequences. The output of the RBI assessment by expert review is in general not repeatable. This is largely due to the fact that it relies on human decision making rather than computer algorithms.

Through the application of RBI, process variables and materials of construction are considered to identify the type of damage that can lead to failure, where it may occur, the frequency of inspections that should take place, and appropriate and cost-effective inspection techniques. As a result, items with a high probability of failure and subsequent impact are given a higher priority for inspection than items that are of low impact, allowing for a more rational application of inspection resources. The entire process results in focusing resources on specific assets that are most likely to pose a risk to the facility.

4.7.2.3 RBI Workflow

The basic steps for the development of an RBI risk assessment and program are illustrated in Figure 4.8, which shows RBI workflow.

Likelihood of Failure

For each facility equipment case, the first step in the process is to assess the likelihood of failure. This is aimed at estimating the likelihood of loss of containment that may result in an incident, which is the result of a specific deterioration mechanism, such as corrosion. The likelihood of failure

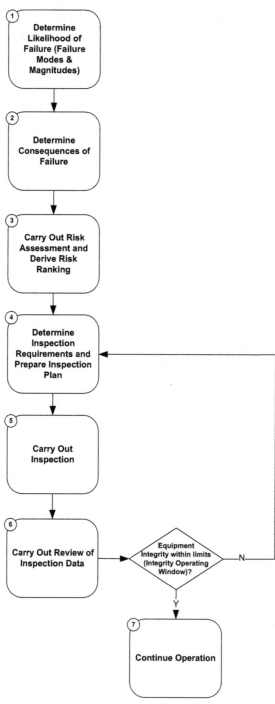

Figure 4.8 *RBI workflow.*

assessment is based on a comprehensive set of predetermined deterioration mechanisms, usually grouped into four generic modes of degradation: wall thinning, environmental, mechanical and metallurgical mechanisms.

In order to ensure that the results of the likelihood-of-failure assessment are reliable, it is important that each possible failure mode is applied to each item of equipment and system in the facility or part of the facility that is under review in the RBI failure mode database, which removes any ambiguity in the assessment.

Consequence of Failure

The consequence of failure assessments employ a hazard consequence rating based on a set of predetermined criteria, which usually comprise safety, business and environmental consequences. The assessment is kept as simple as possible and is based on the most severe but realistic event occurring.

Risk Assessment and Risk Ranking

A risk assessment is performed by using a 5 × 5 risk assessment matrix as shown in Figure 3.2, plotting likelihood and consequence ratings which produces a risk ranking metric. All of the equipment and systems reviewed will have risk rankings assigned. The items with the highest risk ranking will have increased inspection frequencies and tailored plans to suit the specific situations.

Inspection Planning

One of the key outputs of an RBI program is a set of tailored inspection plans for all facility equipment. These inspection plans address the specific risks associated with equipment failure modes. In effect they prioritize the high-risk equipment and assign valuable facility resources accordingly. The inspection plans typically comprise the following: inspection tasks and details, frequency of inspection and due dates.

Once the inspection planning is completed and approved by the FI&R management team, the inspection plans are uploaded and tracked in a computer management tool. This may be an ICMS or CMMS. Results of the inspections are retrieved and reviewed.

Review of Inspection Results

An important aspect of RBI is the review process of the inspection results. The review may take the form of the expert review as described in section 4.7.2.2.

After the initial RBI program is completed and the inspection data is uploaded into the RBI software system, a continuous improvement cycle is created to ensure the system is sustainable in light of new risks and changes to the operation coming to light.

4.8 RELATIVE EQUIPMENT CRITICALITY

The heart of both integrity and reliability is the effective management of risks through use of a prioritization process called *criticality*. Criticality gets to the bottom of what facility equipment and systems are critical to the business objectives, which includes safety, environmental, availability, quality, and overhead cost, among other criteria. It identifies the facility equipment and systems that are a lower priority and therefore require less effort to maintain and operate. Criticality enables the appropriate operating and maintenance strategy to be assigned for facility equipment.

Criticality is a central concept to the FIEM and involves a risk assessment for all of the equipment and systems on a facility. The risk assessment is determined by a number of criteria, including the operating risk from equipment failure. Equipment that will prevent the facility from operating if it fails is identified as "critical." Critical equipment receives greater effort with regards to levels of maintenance and operator care in order to maximize its reliability. More rigor needs to be applied to the design, procurement, construction, and operation of critical equipment and systems.

Ultimately a criticality assessment at a particular facility will rate equipment on a risk priority scale, the deliverable being a criticality risk ranking for each and every item of equipment at the facility. We may refer back to Figure 1.3 showing risk versus percentage of facility equipment, which tells us that our criticality assessment will identify approximately 20% of the equipment as being classified as critical. Once our criticality assessment is complete we can tailor maintenance, operations and inspection strategies to suit based on levels of criticality.

We may calculate criticality of facility equipment as follows:

Relative Equipment Criticality = Business Impact × Likelihood of Failure

Both business impact and likelihood of failure are determined based on a set of predefined criteria that may be tailored for different facilities, since their specific operating conditions and associated risks may differ.

4.8.1 Business Impact

The business impact of a failure can be determined in a number of ways. Typically impact with regards to safety, environment, production, repair cost and repair time are considered. Once the parameters are identified, they are broken down into five levels of severity.

Let us consider repair time as an example. We may consider a worst-case severity ranking of 5 to be defined as the equipment being out of service for greater than 1 week. This is the estimated time until the facility production could be reestablished. A severity ranking of 1 may be defined as less than 4 hours to reestablish production, and so on.

4.8.2 Probability or Likelihood of Failure

The likelihood of failure is determined during the defined operating time of the equipment and is also broken down into five (5) levels of severity. A typical probability or likelihood severity scale can be seen in Table 3.1, "Typical probability or likelihood scale for a qualitative risk analysis."

4.8.3 Criticality Assessment

A criticality assessment may be considered at different levels, which include the facility as a whole, a facility operating unit or at an equipment level. The approach is based on the level of detail required. Criticality assessments should be carried out by a competent, multidisciplinary team, which includes delegates from operations, maintenance, engineering, inspection and reliability. The team should have in-depth experience and knowledge of the equipment or systems being assessed.

Severities of the business impact of the worst-case failure are assigned based on predefined parameters that are developed by the criticality team. These individual severities are normally combined to give a weighted severity rating for the equipment being considered.

4.8.4 Facility Critical Work Flow

The criticality workflow as presented in Figure 4.9 starts with the assembly of the criticality assessment team. The team should be an experienced and competent multidisciplined team so that a balanced approach is taken during the assessment process. The team's first task is to define the assessment parameters that will be used during the criticality assessment. As we have established, the parameters which comprise business impact and likelihood of occurrence need to be tailored to specific facilities, since operating conditions and associated risks may vary. The equipment population is identified for the review in the third step of the workflow.

The criticality assessment team is then tasked with reviewing each item of equipment and assigning a severity rating based on their collective knowledge and experience. Ultimately the outcome of the assessment will produce an equipment criticality list. The facility equipment is now

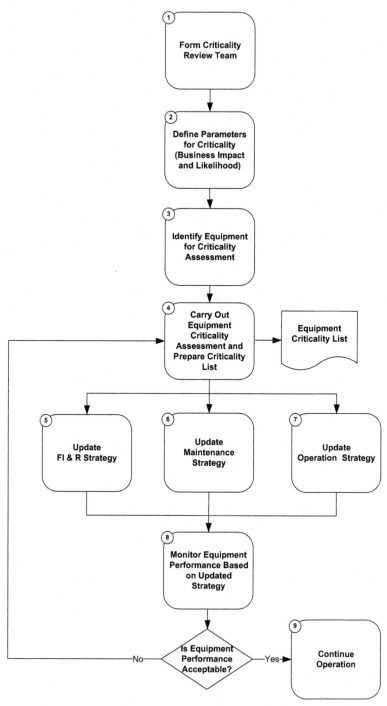

Figure 4.9 *Criticality workflow.*

prioritized in terms of business impact and likelihood of failure. This is an important step forward towards the FIEM. Once we have derived facility equipment criticality, we are able to update our strategy for inspection, maintenance and operations to prioritize our efforts.

4.9 APPLICATION OF CRITICALITY

Why go through all this trouble to assign criticality?

Criticality is a fundamental part of the FIEM. It enables us to put our facility in perspective in terms of what is important. It enables us to prioritize our resources, to optimize our maintenance, operations and facility integrity and reliability strategies. Some of the key applications are noted as follows:

- Prioritization of work orders;
- Prioritization of expenditure requests;
- Determination of risk mitigation strategies for equipment, for example installation of condition monitoring systems or redesign efforts for high criticality equipment;
- Determination of the spare part holdings strategy for the equipment;
- Assignment of FI&R requirements;
- Input into the capital program: for example, high criticality equipment is prioritized for upgrade or replacement;
- Guiding the FI&R group to focus efforts on the most critical equipment;
- Prioritization of root cause failure analyses.

The criticality ranking has a direct influence on the facility maintenance program strategies for equipment. For example, a high criticality ranking will drive the maintenance strategy to hold critical spare parts, apply a preventive maintenance program and potentially have a contingency plan considered in the event of failure.

A low criticality ranking, on the other hand, may result in a maintenance strategy with corrective maintenance only. This means that the maintenance strategy purposefully decides to run the equipment to failure, therefore expending no resource on the equipment until failure, where the equipment is likely replaced.

It is also noted that criticality is applied to equipment in its current condition. Over the equipment life, its condition will inevitably deteriorate. The likelihood of equipment failure should be updated based on an assessment of the current equipment condition. Therefore, the equipment criticality ranking will also change. The facility criticality list is a living document and needs to be reviewed and updated on a periodic basis.

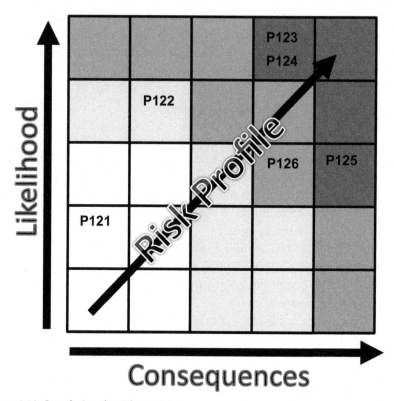

Figure 4.10 *Populating the risk matrix.*

The critical equipment list enables us to plot facility equipment on a risk matrix similar to the one presented in Figure 4.10. On this matrix, you can see example equipment numbers, e.g. P120, P121, P122, P123, P124, and P125, which represent the criticality of a number of facility pumps. On this matrix it is very easy to see the power behind the criticality concept. P123, P124 and P125 should be the focus areas, as these are the high criticality equipment.

Does the pump in the photo shown in Figure 4.11 look like a critical pump?

Yes, the pump in Figure 4.11 is a critical pump, but to look at it you would be forgiven for thinking that it is unimportant. Figure 4.12 is also a critical pump. This has clearly been identified at site. The pump plinth is painted red to symbolize the high criticality rating.

Critical equipment should be clearly identified at site to ensure this concept is engrained into the organization, as we have discussed in our section on risk-centered culture.

Figure 4.11 *An unassuming critical pump.*

Figure 4.12 *A critical pump.*

4.10 ROOT CAUSE ANALYSIS

"All you need to know about improving your facility is encompassed in your failures."

Addressing the root cause of poor performance and equipment failures is of crucial importance. We can make substantial savings in cost and resources by eliminating the "bad actor" equipment items that are routinely failing. Treating the symptoms of poorly performing equipment without really understanding the root cause may produce some short-term gains, but over time it will absorb an inordinate amount of time and effort and could even allow more serious underlying problems to arise. Only when we understand the root causes are we able to take appropriate action to resolve the failure.

For the FIEM to be sustainable, the FI&R strategy must be able to solve problems and adopt a continuous improvement mindset. This is a strategic activity and if it is not addressed we will see deterioration in equipment performance and lower facility production rates. A problem is only stopped from reoccurring if it is eliminated. By focusing on the root cause of the problem, we can be assured that the failure will not reoccur.

A common failing in root cause analysis is that, once the RCA is completed and corrective actions or recommendations are made, the effort to implement them is not followed through effectively. It is essential that the actions arising from any RCA are implemented. Another key point here is that in some cases RCAs will result in some sort of change in the facility. As we have established in Chapter 1, change management is a key part of the FIEM and any change needs to be properly assessed as part of the management of change (MOC) process. We shall review MOC during Chapter 7. Furthermore, if there are similar equipment items that exhibit common failings or malperformance, the results of the RCA need to be communicated throughout the facility organisation so synergies can be leveraged, especially when designing new facilities.

There are numerous methodologies available for RCA. They should follow a systematic approach and focus on examination of the physical evidence to identify the root cause of failure. A typical RCA process should consider the following key components:
- Problem definition;
- Data gathering;
- Problem analysis;
- Cause and effect analysis;

- Root cause identification;
- Corrective actions;
- Report generation.

When carrying out an RCA, a competent multidisciplinary team should be assembled which includes representation from the operations, engineering and maintenance groups at the facility.

The benefits of RCA are immediate with the permanent removal of an equipment failure: in effect the failure disappears for the rest of the equipment's life. Imagine how many manhours and how much money are wasted in dealing with repeated failures. The RCA concept is a critical driver for the FIEM and should be mandated to ensure the facility organization is accountable to address all failures at the facility.

RCAs take time and we have limited resources to carry out all of our failure investigations. Carrying out RCAs effectively requires management commitment and expending resources. This is clearly a constraint along the way as we drive towards the FIEM. We can't afford to overload our overheads with resources to spend all their time carrying out failure investigations. In short we need to prioritize our efforts again when we address our equipment failures. We can carry out RCAs at the right level and effort to reflect the magnitude of the equipment failure.

As part of the FIEM, we can consider three types of RCA investigation, which are aligned to a predetermined failure investigation matrix. This matrix is specific to the facility operating conditions and comprises a scale for outage and scales for cost and duration of breakdown. These criteria determine the efforts needed in the failure investigation.

First, a formal RCA investigation can be employed for major equipment failures. This may be as much as a 2- to 5-day investigation and requires the involvement of the facility management team. It is a formal process that is designed to deal with major failures.

The second type is a "mini" RCA investigation. This is a scaled-down version of the formal RCA and does not require a formal team structure; it may take 1 to 2 days to complete the investigation.

The third is a "5-Why" failure investigation. This investigation is purposefully designed to be quick and easy to execute. It is typically carried out by a mechanical technician, representing maintenance, and an operator, and may be completed in 30 minutes. An example of a 5-Why investigation report is presented in Figure 4.13.

The matrix in Figure 4.14 shows when each of the three types of RCA investigation is used.

5 WHY INVESTIGATION REPORT

Failure Reference No:		Date of Failure:	

Brief Description of Failure:

Background:

Evidence:

Brainstrom Possible Failure Causes:	Tick one probable cause:
	☐
	☐
	☐
	☐
	☐

Ask WHY? Five Times of the Most Probable Cause (Ticked Item Above):

WHY?	
WHY?	
WHY?	
WHY?	
WHY?	
WHY?	

Follow up Actions:

Feedback Required:	YES ☐	NO ☐

Reviewed by:

Figure 4.13 *"5-Why" failure investigation report.*

Failure Investigation Matrix				
	Repair Cost ($ Thousands)			
	Under 10	Between 10 and 50	Between 50 and 100	Over 100
% of Facility Affected	Duration of Outage (Hours)			
	Under 5	Between 5 and 24	Between 24 and 48	Over 48
High Total Facility Outage	Mini RCA	Mini RCA	Full RCA	Full RCA
Significant Facility Outage	5 Why	Mini RCA	Mini RCA	Full RCA
Medium Facility Outage	5 Why	5 Why	Mini RCA	Mini RCA
Low Facility Outage	5 Why	5 Why	5 Why	5 Why

Figure 4.14 *Failure investigation matrix.*

By prioritization of equipment failures, we are able to save time and money in the application of root cause analysis while enjoying the benefit of a failure-free environment.

4.10.1 Cause and Effect

Cause and effect is a popular tool that is used during RCA investigations. It is sometimes called a *fishbone diagram*. An example of a cause and effect diagram is shown in Figure 4.15 for a leaking pump shaft seal. The cause and effect identifies the problem on the right-hand side of the diagram,

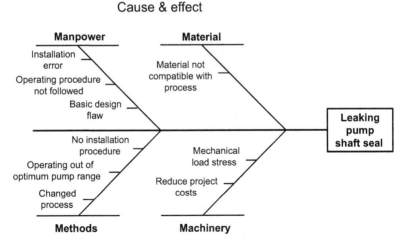

Figure 4.15 *Cause and effect diagram.*

the "effect." On the left-hand side it includes potential causes of the effect. These causes are categorized as:

- Manpower;
- Materials;
- Machinery;
- Methods.

The intention of this arrangement is to structure the way of thinking of the RCA team so that there is a more rounded thought process and a better quality RCA report as a result.

CHAPTER 5

Maintenance Management

Contents

Facility Integrity Management
http://dx.doi.org/10.1016/B978-0-12-801764-7.00005-X

5.1 INTRODUCTION

Maintenance is concerned with quickly correcting failures which are driven by a natural law of system changes. The ultimate aim of maintenance management is twofold: first to ensure that the facility is able to operate as intended and when required – in other words to maximize equipment and systems availability – and, second, to ensure that maintenance resources are optimized (Figure 5.1).

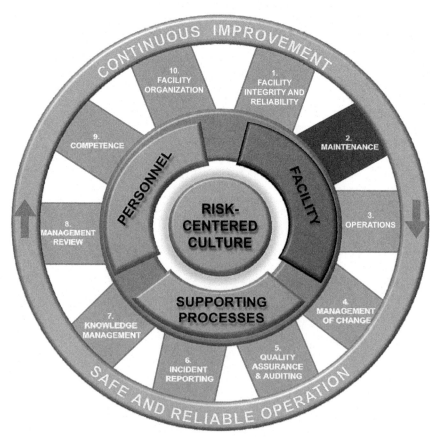

Figure 5.1 *The facility: maintenance.*

There are numerous maintenance strategies available that are employed throughout the industry; however, their associated costs and effectiveness vary hugely. In order to have effective control of facility maintenance, there needs to be a concerted effort to select the right strategy and balance the appropriate deployment of resources.

An ill-defined and poorly implemented maintenance strategy will by nature have little success in delivering the business objectives. If our strategy is based on corrective maintenance, we would see our operation stuck in a "reactive zone" with a high number of common repetitive equipment failures and a higher proportion of equipment out of service resulting in lost production, higher maintenance costs and in general an uncertainty in figuring out what to do next. Reactive maintenance inhibits our ability to meet production targets, absorbs valuable resources and prevents them from being deployed effectively, and in general costs inordinate amounts of money. Furthermore, a reactive maintenance strategy is also likely to contribute to a higher probability of a safety incident.

It is a common perception in the industry for many companies to see maintenance as a necessary burden. These companies have not stayed in touch with world-class maintenance practice and not had opportunity to take advantage of the benefits associated with an integrated and proactive approach to maintenance, facility integrity and reliability (FI&R) and operations such as the Facility Integrity Excellence Model (FIEM). To change this, one of the first steps is to focus on changing the facility culture, moving towards a risk-centered culture as we have discussed in Chapter 3. Maintenance should be viewed as a cooperative partnership with the other facility teams that can significantly contribute to profitability. Commitments will need to be made in order to train and educate the maintenance teams to embrace a risk-based approach and a mindset that encourages cooperative working with the other facility teams. In order to support the new proactive approach, it will be necessary to upgrade maintenance systems and processes, perhaps moving to an up-to-date computerized maintenance management system that interfaces with the other facility groups.

The integrity and reliability of our facility equipment and systems play a huge part in our overall plant performance. The Facility Integrity Excellence Model puts great emphasis on the integration of integrity, reliability and maintenance. As part of this proactive approach, there will be much more involved relationships with maintenance, operations and FI&R groups. Further to this, since facility maintenance, operations, integrity and reliability functions have such inter-relations in each other's territory, it is important that each should know about the others' specific roles, responsibilities, and tools.

5.2 EVOLUTION OF MAINTENANCE METHODS AND SYSTEMS

The maintenance function has evolved significantly over the last 50 years. It is no longer just a body of tradespeople. In the first instance the term *maintenance* was typically perceived as having negative inferences, commonly being thought of as the need to attend to and repair broken-down equipment, an expensive cost center and a huge drain on facility resources.

It has shifted from the all-too-familiar reactive "fix it when it breaks" approach to a proactive and data-driven entity that includes engineers and planners to coordinate the maintenance. If we refer to Figure 5.2 we can see a rundown of the approach to maintenance evolving over the years. The early generation of maintenance mostly involved corrective maintenance in response to equipment failure with basic maintenance routines [5.1]. We have mentioned in the previous section that this is the "reactive zone," which is a hugely inefficient and expensive approach.

With the appreciation of the benefits behind planned maintenance, the approach moved on to time-based maintenance, which involves planning maintenance routines and intervention activities over a prescribed equipment timeline. With the development of new technology and processes, shortly to follow was the introduction of predictive maintenance methods.

With these new maintenance ideologies and technologies came new maintenance strategies to optimize the way we approach maintenance. Focusing on addressing equipment and systems failures or defect elimination was the next big focus. One of the most significant maintenance strategies today that adopted this idea and factored in the maintenance ideologies and technologies was *reliability-centered maintenance* (RCM), which was developed by F. Stanley Nowlan and Howard F. Heap in 1978 [5.2]. RCM was initially developed for the aviation industry and was rapidly adopted by other industries. These new strategies provided a structured process for the application of maintenance tools and technologies activities.

As we progress along the evolutionary timeline of maintenance, we find that most recently we are starting to view maintenance in an integrated team along with the operations and FI&R groups. Collectively, the focus is on identifying and mitigating risk through a risk-centered culture. Ideas such as risk-based methods, including risk-based inspection (RBI) and

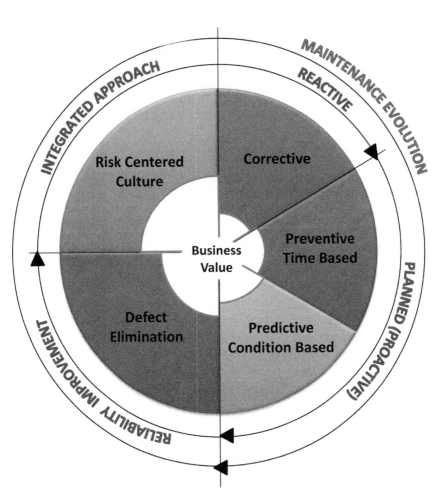

Figure 5.2 *Evolution of maintenance methods.*

RCM, have helped to cement this approach. With this approach in mind, maintenance can be seen in a positive light, as a profit center instead of an expensive cost center. The maintenance profit center recognizes the savings that can be made by adopting a suitable maintenance strategy and prioritizing maintenance resources and activities for the highest risk areas. High performance and low-cost maintenance go hand in hand. The best maintenance organizations exhibit similar characteristics to a high-performing facility and a low-cost maintenance organization.

Liliane Pintelon and Alejandro Parodi-Herz put forward a similar idea of maintenance progression over the years and they suggest that there were four generations of maintenance that have evolved. They suggested that in the 1940s maintenance followed a basic routine approach, mostly corrective maintenance. This evolved considerably to the most recent generation, in which maintenance is a considered to be a cooperative partnership [5.1].

5.3 COMMON FAILURES IN FACILITY MAINTENANCE MANAGEMENT

Before we move into the theory and practical application of maintenance management principles, let's take a moment to appreciate some of the common maintenance management issues and failings that crop up time and time again.

- "Firefighting mode": Despite the focus on preventive and proactive maintenance, many companies find themselves in the reactive zone. In this zone there is a continual battle to "put out fires" moving from one equipment breakdown to the next. In this mode there is no time to be proactive, to resolve the root causes of failures or to train the workforce. An inordinate amount of time is absorbed in this self-perpetuating cycle; there seems to be no time to do anything other than respond to emergencies. This cycle will only get worse as time goes on unless there is a major intervention.
- Repetitive equipment failures: The same equipment items are failing time and time again and our efforts are consumed in dealing with these "bad actors." The focus is on production only, quickly returning the facility back to operating condition with no time to ask why and address the root cause of the failure in order to fix the problem for good.
- Limited resources to perform: The maintenance budget is inadequate and available resources are too little to perform the maintenance work required. This is an all too familiar situation. Maintenance budgets are generally the first to be cut when facility management is under pressure to reduce overhead.
- Poor equipment history: We have no history or a scattered performance and maintenance history of facility equipment at best. This can be a result of many factors, which include maintenance and operator discipline, maintenance and operator training, inadequate systems and processes to support our maintenance program, among many others. A robust and detailed equipment performance and maintenance history is imperative in support of a world-class maintenance program.
- Silo facility groups – operations and FI&R and maintenance: "Us and them"… maintenance teams have limited interaction with operations,

who have limited interaction with FI&R. We are all too familiar with silo teams within facility organizations. This is a highly inefficient and ineffective way of working.

- Underutilized predictive maintenance technologies: "Our maintenance organization is stuck in the past; we have very little exposure to predictive maintenance technologies." One of the major concerns in the industry is expense. There is a perception that predictive maintenance is overkill and not worth considering due to high cost, but huge benefits can be gained from predictive maintenance. Many of the tools and technologies for predictive maintenance have been around for a number of decades and are relatively inexpensive today. While the application of predictive technologies can be expensive, we can manage this by proper application of maintenance strategies and targeted selection and deployment of predictive technology. We are likely to see that the cost/benefit ratio will be in favor of predictive tools and technologies.
- Narrow focus only on production: "Production, production, production…" Facility management focus is entirely on production, while the other facility functional groups including maintenance and FI&R are neglected and suffering.

This snapshot of some of the common failings in facility maintenance management may look familiar. These failings have the potential to manifest themselves into identifiable and tangible obstructions such as:

- Decreased production;
- Low morale;
- Low mean time between failure (MTBF);
- Increased maintenance costs;
- Increased spare parts inventory and cost;
- Reduced product quality;
- Poor margins.

Many of these failings are due to maintenance organizations being stuck in the reactive zone. The common complaint is that these organizations claim to have no time to change, no time to implement a risk-based maintenance strategy such as reliability-centered maintenance. Clearly there will be an upfront investment required and an effort to change, but once the program is in place and established, it will actually improve the efficiency and effectiveness of facility maintenance and free up time rather than consume it. There is a significant opportunity to reduce maintenance costs and free up resources and in parallel improve MTBF and MTTR.

5.4 MAINTENANCE MANAGEMENT CONCEPTS

It is common practice to categorize maintenance management into two main categories, as follows:

1. Reactive Maintenance. This refers to breakdown maintenance and corrective maintenance. Reactive maintenance was one of the first widely used maintenance strategies in the industry. Reactive maintenance is performed usually as a result of equipment failures in response to unplanned downtime.

2. Proactive maintenance. This type of maintenance is performed before the necessity of the situation requires it. It is performed to detect and correct conditions that could lead to equipment degradation (rotating and static) before the cost of doing so increases. Proactive maintenance improves the facility equipment through improved engineering design, workmanship, planning and scheduling. Proactive maintenance may be either preventive maintenance or predictive maintenance.

   a. Preventive maintenance is interval based, which means that maintenance activities are performed in line with a predefined maintenance schedule. Preventive maintenance tasks may include inspection, replacement, and servicing tasks that have been scheduled in the future in order to ensure the equipment remains functional and fit for purpose.

   b. Predictive maintenance is a systematic approach to maintenance to determine the equipment condition, whether equipment is near failure and therefore needs to be replaced or repaired. In doing so, more expensive unscheduled (reactive) maintenance can be avoided. Predictive maintenance activities are associated with preventing costly major repairs or unscheduled downtime. Predictive maintenance employs a number of diagnostic and monitoring tools and measurement systems.

Proactive maintenance should only be performed when it is cost effective to do so.

Facility maintenance managers may also choose to run items of equipment until they fail. As strange as it sounds, it is an industry-accepted strategy and is usually driven by economic factors. This strategy is known as "run to failure" (RTF) and is commonplace in an RBM program that focuses on prioritizing maintenance resources such as RCM.

The diagram in Figure 5.3 shows the various maintenance strategies.

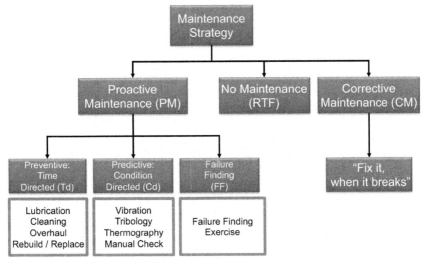

Figure 5.3 *Classification of maintenance.*

5.4.1 Maintenance Repair Time

When we talk about maintenance repair time, it is often useful to break this down and understand the component parts involved. The illustration in Figure 5.4 helps to explain the typical make-up of maintenance repair time. The illustration demonstrates that the actual time to carry out a maintenance repair task may only represent a proportion of the total downtime. The total downtime may be made up of a number of events.

As an incipient failure starts to manifest itself to the point of detection (Figure 4.3: PF curve – equipment failure), we may choose to operate the facility with a reduced output. This is an attempt to lessen the load and hence the stress on the failing equipment and avoid a catastrophic failure. During this time when maintenance spare parts are ordered for the equipment, we may face a supply chain delay, which includes supply chain administration delays, logistics time and procurement lead time. Once the equipment spare parts arrive, we are almost ready to execute the maintenance repair or replacement work. In the process we may also be faced with maintenance delays which comprise administration lead time along with delays associated with mobilizing maintenance resources.

The equipment is now taken offline and associated production is reduced to zero until the maintenance task is completed. Access needs to be provided, which includes preparing the required paperwork, a job plan and a risk assessment. We can then apply for a work permit and make the

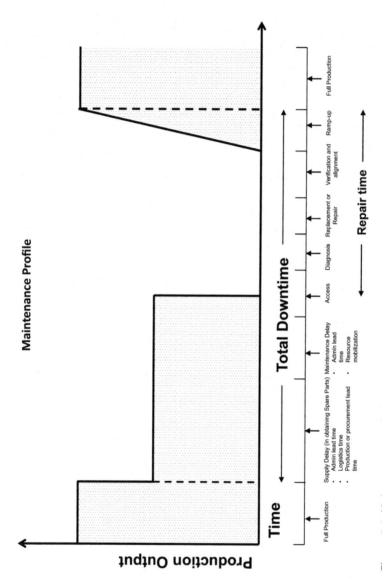

Figure 5.4 *Maintenance repair time.*

equipment and local area safe to work by carrying out electrical and mechanical isolations, for example; all these activities add to the delays.

The maintenance team can now mobilize to the job site to diagnosis the failure in detail and gather the necessary evidence to establish the root cause of the failure. The team can also carry out the necessary repairs or replacement work to restore the equipment. This may also include alignment activities and finally a quality check is performed to ensure the repair or replacement meets the high level of quality set by the FIEM. The maintenance history and work order can then be signed off by a maintenance supervisor and the equipment handed back over to the operations team for power up and testing. Once the equipment is satisfactorily tested, production is ramped up to the desired production capacity.

The key point here is that there are many hidden activities and potential delays that need to be considered when either equipment fails and maintenance needs to be carried out in an unplanned manner, and even when maintenance is carried out in a planned manner. When we talk about maintenance repair time, we need to be aware of exactly what these hidden activities and potential delays may be so that we can be more effective in minimizing MTTR and increasing uptime.

5.4.2 Corrective (Reactive) Maintenance

Corrective maintenance (CM) involves the replacement or repair of equipment after it fails. In response to equipment failure, CM tasks identify the failure (it may be an equipment component or equipment item) and rectify the failure so that the equipment can be reinstated and the facility production restored. CM tasks are prioritized so that the high-priority tasks that may be safety related or affecting production are addressed first.

CM is in general low cost because it can generally be performed with a fewer number of resources and maintenance infrastructure, including tools, technologies and expertise. The consequence, however, is that it is inefficient and in the long term it can be very expensive because failures generally result in catastrophic events, which means there is more damage that needs to be repaired and hence the MTTRs are longer. CM also does not focus on the root cause of the equipment failure and therefore MTBF will be much lower than with proactive maintenance. In other words, there will be many repeat failures.

5.4.3 Proactive Maintenance

Proactive maintenance is determined by equipment criticality. Criticality is synonymous with the most successful maintenance strategies and enables valuable maintenance resources to be assigned efficiently and appropriately. Proactive maintenance determines the depth and intensity of maintenance activities and provides a guideline for work prioritization for the equipment. It involves the tailoring and optimization of maintenance techniques and technologies to meet the requirements of each equipment item or system. This requires a dedicated approach by all of the facility groups and thus encourages cooperation between the different facility functions since it involves aspects of each.

It can be categorized further into preventive maintenance, which is time-based or "time directed" (Td), and predictive maintenance, which is a "condition directed" (Cd) approach. This allows effective utilization of resources and focuses the facility organization to adopt a problem-solving mindset, particularly in the prevention of reoccurring equipment failures.

Proactive maintenance also in some cases makes a deliberate decision to run equipment to failure. This is considered in circumstances where there is an economic advantage to do so. "Run To Failure" or RTF is exercised under very controlled circumstances, particularly when implementing a reliability-centered maintenance (RCM) program.

5.4.4 Preventive Maintenance

The preventive maintenance (PM) strategy was one of the very first maintenance strategies and it is still an effective one. It is also referred to as time directed and comprises two types of maintenance activities: nonintervention type activities, which include monitoring and inspection of equipment, and intervention type activities, which include repair or replacement.

Nonintervention tasks are related to monitoring and inspection activities such as maintenance rounds. A maintenance round is a predefined route on a particular area of the facility, containing a well-defined number of equipment items and systems that need to be monitored or inspected. It is the maintenance team's job to complete their round during each shift, assessing each and every item of equipment in terms of performance and signs of performance degradation or *variances* to their operating norm. The output of each round will include a written record of the condition of equipment that is the topic of discussion during the production meeting and maintenance shift change-over meeting.

Nonintervention tasks may also include scheduled statutory inspections in line with government legislation, including lifting equipment such as hoists and pressure equipment such as relief valves.

Intervention maintenance tasks are time-based tasks and may include repair, replacement or intrusive servicing of equipment. Intrusive servicing of equipment usually requires a shutdown of equipment and may necessitate the equipment to be dismantled. While the servicing is done, it is a good time to inspect for evidence of incipient failures. During this process if failure evidence is found it can be addressed immediately.

Sometimes when it is not possible to repair or replace equipment because it is operating, for example, a number of preventive maintenance tasks are grouped together and executed during a preplanned facility shutdown. A preventive maintenance plan can allow the facility to schedule production shutdowns for repairs, inspection and maintenance in this way.

There are a number of prerequisite processes and systems that a preventive maintenance program requires in order to be effective:

- Maintenance management system in order to plan and schedule maintenance work;
- A discipline among the maintenance team to ensure maintenance routines are completed and closed out properly;
- An up-to-date record of equipment failures and repair history;
- An appreciation of equipment failure history and feedback loop to adjust PM task frequencies.

A well-developed and run preventive maintenance strategy can be effective at preventing equipment failures to a large extent. It can provide a tool for equipment repairs to be planned and orderly. One of the side effects of a preventive maintenance program is that it can be wasteful with regards to utilization of resources and therefore expensive in the long run.

5.4.5 Predictive Maintenance

Predictive maintenance is a very powerful maintenance strategy. It involves monitoring for evidence of abnormal operation or *variances* within the equipment. The extent of the variance and the rate of change to normal operation is tracked and used to predict the time of failure. As mentioned already, predictive maintenance is also referred to as condition directed (Cd) and is based on the concept that each item of equipment follows a failure cycle, as we have presented previously in Figure 4.3.

Predictive maintenance focuses on identifying a failure as early as possible along the P-F interval. The earlier a failure is detected, the more time there is to decide how to manage the equipment and balance the requirement for continued operation.

There are many Cd tools and techniques at the disposal of the FI&R and maintenance groups. These tools and techniques all focus on detection of equipment variances and measure the rate of change of variances.

This is in order to be able to predict future equipment performance and to make informed judgements on the next course of action. The recommended action is usually based on equipment criticality, the variance deterioration versus the operating envelope (OE) and forecasted trend analysis.

Predictive maintenance techniques also overlap the facility operations group in that technology permits online monitoring of equipment condition, which tends to be expensive and its application is very selective. There are now new advances in online monitoring technology that include wireless technology to reduce the cost of installation of hardwired sensors. Furthermore, web browsing of online equipment condition data and an alert system to inform the operator or maintenance team of any variances by email or text messaging are available. Some of the more pertinent Cd techniques are noted in the following subsections.

5.4.5.1 Vibration Monitoring

Vibration monitoring provides a good quality source of information about the health condition of facility equipment. It is a core element of any predictive maintenance condition monitoring effort in a facility. Typically a member of the predictive maintenance team, a vibration monitoring technician would follow a predetermined maintenance round in the facility and record vibration readings using an electronic handheld device. The predetermined maintenance round would include markers whereby the electronic handheld device can be accurately placed to measure equipment vibration. For example, these may be on the X and Y axis of a pump bearing housing. The data is then saved on the handheld device and trended. A software package can then diagnose the vibration data and identify if it exceeds the operating envelope and, if so, an alarm is activated. There are many different vibration monitoring techniques available that increase in cost as the accuracy and number of features increase.

These data are particularly useful for the identification of incipient equipment failure modes, which can give confidence in operating the equipment. It can also be used to check that equipment servicing has been done correctly.

5.4.5.2 Thermal Techniques

There are a number of thermal measurement techniques available for equipment condition monitoring. Some of the leading techniques are noted in the following paragraphs.

Thermography is used to measure thermal (infrared) energy emitted from an object. Infrared energy is similar to visible light but not detectible by the naked human eye. Instead a thermal camera and software is used, which can provide a multitude of equipment condition data.

Thermography can be a very useful tool and suitable to a number of condition monitoring applications for a variety of equipment, both static and rotating. These include inspection of and identification of problems in electrical equipment, facility refractory, heaters, motors and rotating equipment, among others.

Thermal point measurement measures the temperature of a specific sensor that is in contact with an equipment item surface. Dedicated temperature sensors or thermocouples are strategically fixed to points of concern on facility equipment, such as a pump or motor bearing. The sensor measures temperature directly. The data can be integrated into an online condition monitoring system and trended. Temperature handheld devices or handheld pyrometers are also available which can measure the emitted infrared radiation from a surface. This is read locally to the equipment during an operation or maintenance round.

5.4.5.3 Oil Analysis

By operating facility equipment with degraded or contaminated oil and hydraulic fluids, we will see accelerated wear, higher energy costs and eventually premature equipment failure.

The effectiveness of oil lubrication can be determined by analyzing the level of oil degradation and water and debris content in the oil. This is an important condition monitoring technique because it provides a proactive indication of the onset of equipment failure. Ultimately it allows us to improve the reliability of our equipment.

An oil analysis condition monitoring program takes oil samples from facility equipment and tests them for oil condition. These are compared against a set of oil performance limits, such as water and debris content. The FI&R team will then make an assessment for each sample to identify if there are any deviations from the oil performance limits. The results are published in an oil analysis report along with corrective actions and communicated to the FI&R, maintenance and operations groups.

5.4.5.4 Online Asset Condition Monitoring

On-line condition monitoring systems are available to measure most equipment condition parameters, where a sensor exists. These can include pressure, vibration, temperature and oil condition. Sensors are fixed to facility equipment at predefined locations where online measurements are taken. The output of the sensor will normally be converted into a 4–20 mA signal and integrated to the existing facility control system.

It is often useful to combine condition monitoring methods to cross reference equipment performance and condition deterioration. Combining condition monitoring techniques tends to provide better and a more reliable indication of equipment condition.

5.4.6 Application of a Condition Monitoring Program

There is a balance with regards to the benefit of a condition monitoring program versus the cost of installation and maintenance. The cost of a condition monitoring program can be offset against the benefits, which will result in a reduction in the extent of FI&R and maintenance work and freeing up of FI&R and maintenance resources. Additionally, a condition monitoring program can provide assurance in the accurate monitoring of equipment performance and the onset of failure.

The assessment of the application of a condition monitoring program should be carried out by competent and experienced facility personnel, who have a good understanding of the RBI or RCM evaluation that gave rise to the recommendation for monitoring.

The selection of the condition monitoring equipment needs to identify the precise equipment and instrumentation that is required for monitoring purposes. There needs to be confidence that the equipment will accurately and reliably monitor the identified failure degradation so that there is clarity as to when the failure has progressed to exceed the integrity limits.

Once the operating performance of the equipment in question approaches the integrity limits set into the condition monitoring program, an alarm is triggered. The intention is for the alarm trigger level to be highlighted to the FI&R and maintenance group to take appropriate action and carry out remedial action. Remedial action may include adjustment of the operation process control, repair or replacement in order to remove the causes of the failure degradation.

In order to get the best out of a condition monitoring program, it is important to commit time and effort to monitor critical equipment failure degradation performance and interpretation of the data. The data should be trended over time so that an appreciation of equipment performance can be gained and so that any increase in the rate of degradation is readily observed in time for appropriate action to be taken.

When the rate of degradation of a failure increases, an evaluation by competent FI&R and maintenance engineers and operators and inspectors should be carried out. The evaluation usually involves a thorough inspection of the equipment at its location in the facility in order to establish the extent of the degradation. A root cause analysis, mini root cause analysis or 5-why investigation is then conducted. Finally, an action plan that is collectively prepared by the FI&R, maintenance and operations team is prepared in order to decide on the next course of action.

5.4.7 Condition Monitoring Strategy Development

Careful consideration needs to be given in the development of a strategy for a condition monitoring program for a number of reasons. First, there is a high upfront capital investment required at the outset. The program also needs to be maintained, which requires maintenance of the condition monitoring equipment as well as ongoing costs for the team of condition monitoring technicians. Subject matter experts (SMEs) also need to be hired for specialist knowledge and interpretation of condition monitoring data that is collected from the facility.

The condition monitoring strategy is often driven by the output of risk-based methods such as risk-based inspection (RBI) or reliability-centered maintenance (RCM). Risk-based methods, as we have discussed in Chapter 4, are underpinned by the understanding of equipment failure modes, which need to be identified and evaluated. In addition to RBM, a condition monitoring strategy requires a criticality assessment to be performed on facility equipment. This is important so that high-priority critical equipment is identified and appropriate measures are taken to monitor its performance, which includes condition monitoring techniques, which further justify the capital expenditure, for example.

Referring to Figure 5.5 we can see the condition monitoring strategy workflow. First we need to develop a suitable condition monitoring scheme for the equipment item or system in question based on equipment criticality and an RBM assessment. The driving parameter for the scheme is the degradation of the identified failure mode, which needs to be monitored. As the failure progresses, the integrity limits are eventually exceeded and therefore intervention such as maintenance activity is triggered. Once the condition monitoring scheme is established for facility equipment, it is important to set the operating envelope as indicated in the second step of the workflow. OEs (operating envelopes) are effectively the operating limitations that facility equipment and systems are able to safely and effectively operate

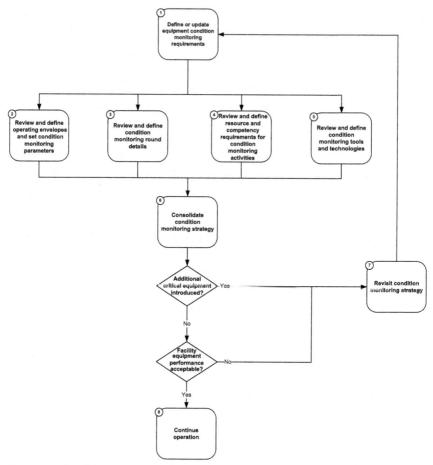

Figure 5.5 *Condition monitoring strategy workflow.*

within (Chapter 6). If equipment exceeds the predefined operating envelopes, a trigger is highlighted to the FI&R and maintenance group to take
appropriate action. It is then important to develop condition monitoring
rounds for facility FI&R condition monitoring technicians or to incorporate new monitoring activities into existing condition monitoring rounds.

A review of the FI&R condition monitoring resources is essential in order to ensure the condition monitoring program is achievable and sustainable. Furthermore, the condition monitoring team should be sufficiently
trained and experienced in order to ensure the correct data is collected
and that it is collected the correct way. Condition monitoring data must be
"clean"; in other words, it must not be contaminated with spurious results
due to errors in collection of the data.

Once we have set the OEs, updated or introduced new condition monitoring rounds, reviewed and addressed resource deficiencies to execute the condition monitoring strategy, and aligned the appropriate condition monitoring tools and technologies to the condition monitoring program, the condition monitoring strategy is consolidated and recorded. This is important because, in line with the FIEM principles and the Shewhart cycle, we must be able to assess our performance and continually improve. Therefore, as the facility equipment criticality list changes and as our confidence in particular equipment performance develops, the condition monitoring strategy should be revisited and updated accordingly.

5.4.8 Managing Maintenance Backlog

Maintenance backlog is concerned with maintenance activities that have been planned but not yet scheduled to be executed. It can be defined as the number of estimated maintenance man-hours of labor required in order to complete the total number of authorized maintenance work orders, divided by the maintenance manpower available to execute the work orders. It typically excludes nonproductive time such as holidays and sick leave, etc. Maintenance backlog can be expressed in terms of man days or weeks and can be tracked for each maintenance team or crew. A good target for backlog should be between 2 to 6 weeks; however, it is important to have a satisfactory number of maintenance backlog work orders in order to allow time for proper work order planning before the work is actually scheduled to be executed.

Typically a maintenance backlog report is generated in order to appreciate and respond to utilization of maintenance resource constraints or optimization of maintenance team sizes, determination of overtime and assessment of third-party contractor services.

5.4.9 Spares Management

The management and control of spare parts and materials in support of maintenance work orders is a crucial part of any maintenance program. Spare parts tend to carry a high cost which consumes a large part of the maintenance budget.

The challenge of maintenance spares management is to ensure that there is a balance between the minimum number of spares needed to support the facility and cost of carrying the spares. We could have a spare for each and every equipment item in the facility but our costs and business would not be sustainable.

The maintenance group must carry sufficient spares so that in the event of equipment failure the repair or replace time is kept to a minimum and the facility outage is minimized. The quality and availability of equipment spare parts contribute to the equipment's ability to meet production needs and help optimize equipment servicing or overhaul cost. Spares holding stock should be kept to a minimum to reduce the overall cost of spares, both in terms of capital cost and maintenance of spares cost.

Our goal is to optimize maintenance cost while minimizing our safety and environmental risk and our potential for lengthy periods of downtime due to unavailability of spares. We can go about this by alignment of the criticality concept with maintenance spares holding. Maintenance spare parts have an effect on the criticality ranking, particularly of rotating equipment because of the spare parts lead times, which may be very long.

There are many ways to classify equipment maintenance spares holding. In the interest of illustrating the point we can describe three rankings: High, Medium and Low.

- A high ranking may mandate that maintenance spare parts are available within two hours of equipment failure.
- A medium ranking may require maintenance spare parts to be available within 8 hours of equipment failure; these may include spare parts that are required as part of equipment overhaul.
- A low ranking may include equipment spare parts or repaired parts to be available within 5 days.

It is not unusual to have spare parts and material maintained outside of the facility, by a supplier or at a vendor's facility, provided that the maintenance and spares-holding procedure meets the required facility quantity requirements as well as the availability requirements.

Maintenance spares storage needs careful consideration. Spares must be stored in a safe and controlled environment depending on the nature of the spares, so that they cannot be damaged either by mishandling, corrosion, the effects of the sun and humidity, and so forth.

5.4.10 Spare Parts Interchangeability

It is common practice to standardize maintenance spare parts as much as possible. This is important because it means that we can minimize the number of maintenance spare parts that are held as inventory and hence the overall cost. However, this needs to be balanced with equipment criticality.

A spare parts interchangeability register should be completed for facility equipment with spares parts holding. The register holds information about the spares parts associated with each equipment item, including component vendor reference number, cost, details of materials of construction, etc. This information is recorded and uploaded onto a central maintenance database. The spare parts interchangeability register accurately details common spares holding, which may be extensive in some cases, such as consumable spares including gaskets and bolts.

Efforts need to be made in identifying slow-moving stock maintenance spares. In general if equipment spares are not utilized after 3 years, these spares should be reviewed in terms of their priority and consideration given as to their retention. Slow-moving stock maintenance spares tend to deteriorate after a long shelf life despite actual maintenance activities being performed for spare parts and spare equipment, and therefore close scrutiny needs to be given to the benefit of holding these items. Furthermore, spares may also become obsolete over extended periods of time, which is an important factor in optimizing spares holding.

5.4.11 Maintenance Management System

The term *maintenance management system* (MMS) refers to a system that is used to schedule maintenance, inspection or testing activities. It is a critical element of any maintenance function. It provides a platform to enable the facility organization to satisfy operational, financial, regulatory and safety goals by systematically managing the execution and completion of facility maintenance tasks.

An MMS is one of the primary building blocks for the safe and successful management of facility equipment. It should be viewed as a valuable and central facility process. In addition to scheduling maintenance tasks, an MMS can also enable the effective management of a number of other components, including resource management, production plans, schedules, task plans, cost, spare parts, equipment maintenance and failure history. The primary output, however, should focus on ensuring a well-maintained condition and functionality of facility equipment and systems.

An MMS should be integrated with the other facility functions including FI&R and operations. There are obvious overlaps between the functions; for example, operations and FI&R inspections can be scheduled through the MMS. Equipment history and performance data which is captured in the MMS must be shared, reviewed and acted on by the core FIEM functions including FI&R and Operations.

We may consider the following objectives which should be a target output of an MMS:

- Ensuring that all facility equipment is fit for use and in good working condition;
- Reducing the mean time between failures;
- Reducing the mean time to repair;
- Reducing critical incidents/near miss incidents;
- Increasing maintenance personnel skills/work experience.

An MMS must be successfully integrated in order to achieve the required objectives.

5.4.12 Computerized Maintenance Management System

An essential building block for an effective maintenance management system is a robust and reliable maintenance work order system. A maintenance work order is the primary input to enable the proper planning and scheduling of maintenance activities at a facility. In order to effectively manage the large amount of maintenance work, it is often necessary to employ a computerized maintenance and material management system, or CMMS. A CMMS is an integrated software package that allows expedient and automated management of work orders and management of associated support and supply activities. It is effectively a database that stores and manages a large amount of data, including equipment details, equipment history, spare parts, maintenance resources and materials, and costs, among many other items.

In many cases a CMMS has additional features such as analytical tools that can prepare cost estimates for maintenance work, resource load maintenance work and manage spares holding or inventory at a facility as well. One of the main drivers for a CMMS is to provide simplicity in the management of facility maintenance work.

The main elements of a typical CMMS are as follows:
- Manage maintenance work orders;
- Enable effective planning and scheduling of maintenance work orders;
- Manage maintenance backlog;
- Provide a register of equipment in a facility including criticality;
- Provide a tool to manage maintenance spares and materials;
- Provide a record of maintenance and equipment performance history;
- Cost control and analysis of maintenance (integration with facility financial accounting);
- Produce maintenance reports (flexibility for user-defined reporting).

It is imperative that a CMMS is easy to use and does not hinder the basic principles of maintenance management. The maintenance user interface should be user friendly with clear data entry points and effective reporting. The system should be able to track equipment failure history, downtime and effectively facility availability. This is important because it enables the FI&R group to perform root cause analysis and assess frequency of failures and downtime in order to develop trends and therefore take appropriate action.

5.5 EQUIPMENT MAINTENANCE AND OPERATING PLAN

An Equipment Maintenance and Operating Plan (EMOP) is the primary process used as a part of the FIEM on a facility for defining operating and maintenance strategies. It ties together three of the fundamental elements of facility integrity management: FI&R, operations and maintenance and in the process promotes the values of the risk-centered culture. EMOPs provide baseline information for facility equipment such as operating envelopes for each of the equipment or system operating parameters.

As with many of the FIEM principles and processes, EMOPs are developed based on the assessment of risks. Equipment failure modes are identified and the risk of failure is assessed. The resulting mitigating activities to reduce the risks and prevent or manage the failure are developed. The output of this assessment is a number of mitigating activities that can be performed by the operators, inspectors or maintenance technicians, which are recorded on the EMOP.

The development of EMOPs is done jointly by both by the operations and maintenance teams.

5.5.1 EMOPs and Facility Operations

EMOPs define tasks for facility operators. These tasks are based on ensuring the availability of facility equipment, i.e. prevention of equipment failure. The tasks are developed based on a criticality assessment, which determines the priority equipment and systems. The defined tasks are kept as simple as possible to ensure there is no ambiguity and potential for mistakes in the field.

EMOPs effectively provide a blueprint for the operations teams in terms of consistent standards for inspection and a check that facility equipment is operating effectively and any variances are identified as early as possible.

EMOPs also provide a definitive training syllabus for operators, based on actual facility equipment operating conditions. This is effective for new operators and also refresher training for existing operators, which also provides a paper trail for auditing and continuous improvement.

5.5.2 EMOPs and Facility Maintenance

In terms of the maintenance team, there is commonality in the EMOPs with operations in that EMOPs provide a consistent standard for inspections and checks for maintenance as well as operations. There is also a basis for task-specific maintenance training at the facility for the maintenance team. This is to ensure that maintenance rounds are performed to minimum quality standards and that there is a foundation for continuous improvement and auditing. EMOPs contain detailed maintenance procedures for facility equipment as an output of the maintenance strategy, which we shall discuss in section 5.7.

5.5.3 EMOP Structure

A typical equipment maintenance and operating plan comprises the following critical information:
- Equipment criticality;
- Equipment information: manufacturer, date of manufacture, materials of construction, etc.;
- Operations information: operating strategy, operating envelope, etc.;
- Maintenance Information: lubrication details, filter details, failure history, spare parts details, etc.;
- Equipment maintenance plan;
- Equipment operating plan.

5.5.4 Equipment Maintenance and Operating Card

The FIEM takes equipment maintenance and operating plans one step further and summarizes the critical information on a card. The equipment maintenance and operating card may be prepared for facility equipment with medium to high criticality rankings and is displayed beside the equipment in a weatherproof frame.

An example of an equipment maintenance and operating card (EMOC) is shown in Figure 5.6. The card contains information that is deemed to be important and useful to have quick access on site. It immediately highlights equipment criticality. There are many ways to do this, such as color coding: all highly critical equipment has a red border, medium, orange border, etc. The criticality ranking is also noted on the card. This is important so that all facility personnel are immediately

Equipment Maintenance and Operating Card

Re-circulation Pump			
Equipment Number	P123		
Manufacturer		**Model No.**	
Serial No.		**Criticality Rank**	
Driver Details		**Power Supply**	
Duty			

Materials of Construction	**Casting Cover**	**Impeller**	**Shaft**
	Volute Casing	**Seal Cover**	**Shaft Sleeve**

Process Variables	**Design**	**Normal Operation Range**	
		Minimum	**Maximum**
Pump speed (RPM)			
Suction pressure (Barg)			
Discharge pressure (Barg)			
Lube oil temperature (°C)			
Flow Rate (m^3/hr)			
Bearing vibration level (mm/s)			

Lubrication Specification	
Lubrication Oil Filter Type	
Critical Spares Holding	

MCC Board Details	

What Can Fail?	**Operations and Maintenance Checks**
Lubrication	The oiler is not leaking
	The oil condition is ok (colour is clear and no debris present)
	The oil temperature is not overheating (can be touched)
Driver Seals	The seal flush temperature is not overheating (can be touched)
	The seal flush has a good flow rate (flow meter is positive)
	There is no leakage from seals / casing /gaskets/ connections
Bearings	The bearing housing temperature is not overheating (can be touched)
	The bearing housing is not vibrating excessively
Driver (Motor)	The motor temperature is not overheating (can be touched)
	That there is a good air flow across the motor vanes (feel air flow)
Pump Efficiency	There is no popping / grinding sound present (potentially cavitation)
	There is no abnormal noise
	Ensure all valve positions are in correct positions (as indicated)
General	Gauges - replace any damaged / inaccurate gauges
	Guards - check seal and coupling guards are in place and secure
	Clean bedplate - remove debris and oil accumulation
	Housekeeping - check area for tidy as required
	Labelling - Ensure correct labelling of all system equipment
	Vents / drains - ensure all connections are plugged as necessary

Figure 5.6 *Equipment maintenance and operating card.*

aware of the importance of critical equipment to differentiate it from noncritical equipment.

Referring to Figure 5.6, the equipment maintenance and operating card shows the key process variables for critical equipment. These process variables include the specific operating parameters for equipment and facility process performance, such as pressure, temperature, vibration and flow rate. The operating envelope for each parameter is also shown as a normal operating range. This is important because it is therefore possible to compare the actual operating conditions of the equipment against the operating range; thus the equipment can be easily monitored for any variances or degradation in performance.

Equipment maintenance and operating cards also contain a summary of the key inspection and monitoring activities that must be carried out as part of the operations and maintenance rounds. By making these activities visible for all to see, it is possible for all site-based facility staff to "get involved" and appraise the monitoring efforts themselves, reporting any variances to the operations or maintenance teams.

Equipment maintenance and operating cards provide a visual indicator that contributes to risk-centered culture. They facilitate better risk awareness to the site facility organization, from maintenance technicians to engineers to facility managers.

5.5.5 Surveillance Rounds

Monitoring the facility equipment and systems is done periodically, often during each day or each shift. Maintenance activities that involve walking around the facility equipment and systems are commonly referred to as "rounds." The literature also may refer to "surveillance rounds." A round covers a prescribed route in a facility with a defined list of equipment to monitor. Rounds are performed by all facility groups: operations, maintenance and FI&R. All of these groups monitor equipment for a common purpose but have different factors and different levels of detail.

Facility operators and maintenance technicians monitor their facility for variances and report any findings so that appropriate action can be taken. The variances that are picked up may be recorded as simply as recording a note on a clipboard with a checklist covering specific checks, as presented in Figure 5.6.

Experience suggests that, more often than not, paper-based monitoring data ends up being filed in a dusty cabinet in a corner of the maintenance office and little effort is made to review and communicate the data. It is imperative that any variance management system in use at a facility is able to efficiently and effectively serve the operations and maintenance teams in the prompt recording and acting on equipment variances. We shall review the FIEM unit monitoring principles in section 6.6 in Chapter 6.

Many facility integrity groups are moving to electronic data recording. Devices such as electronic clipboards are more and more often being employed as a means to record and manage maintenance, operations and integrity facility data. One of the key drivers for the increased use of handheld digital devices at facilities is the fact that, over the last few years, costs have come down considerably.

These handheld devices enable equipment data to be recorded in an electronic format. The data can then be taken back to the maintenance or operations control room and downloaded onto a software package. Although set-up and implementation costs are much higher than conventional methods such as paper clipboards, there are clear benefits. It is important, however, not to allow technology to distract from the basic principles of facility integrity management.

A process that consistently delivers quality and robust surveillance rounds is imperative for the FIEM. This drives reliability improvements across the facility, which in turn lowers maintenance costs and improves safety and environmental performance. Surveillance rounds are carried out by all three facility groups: maintenance, operations and FI&R. It is important to understand the differences between the three, as described in the following subsections.

5.5.5.1 Maintenance Surveillance Rounds

Maintenance technicians undertake rounds which tend to follow a prescribed route on a particular geographical area or process area at the facility. The maintenance teams conducting the round would typically cover all equipment in their respective facility areas.

Often the maintenance rounds are broken down to a general daily round, which is less time in order to cover more equipment; typically the round will take 30 minutes or so to complete depending on a number of factors. There may be a more detailed weekly round that may typically take half a day or so to complete. Maintenance rounds tend to focus more on rotating equipment. This is because maintenance technicians tend to have the technical know-how as to the inner workings of rotating equipment and machines, and therefore understand the difference between healthy equipment and equipment that is malperforming.

There may also be a requirement for additional maintenance rounds that are targeted particularly on poorly performing equipment. These bespoke additions may even be added onto the general round.

There is a large amount of information to record each day from maintenance surveillance rounds. This information may be recorded in numerous ways, from hard copy check sheets to handheld electronic devices.

5.5.5.2 Maintenance Round Checklist

It is good practice to prepare a maintenance rounds surveillance checklist. This checklist should detail the basic maintenance checks that are to be performed during the maintenance surveillance round. The checklist is a useful guide to ensure that all the major checks are completed. It is particularly useful for new maintenance team members to ensure thorough rounds are consistently completed. The checklist is also a "living" document that should be reviewed and improved periodically or when changes are made to the facility. An example of a checklist is illustrated in Figure 5.7.

5.5.5.3 Operator Surveillance Rounds

Operator surveillance rounds have a different focus than maintenance surveillance rounds and tend to concentrate at a higher level than the individual facility equipment. They focus on ensuring the *process* is functioning effectively. There is, however, an overlap with maintenance rounds and the operators should also monitor the facility equipment for advanced signs of failure. The operations teams conducting the round would typically cover all equipment in their respective facility areas.

The information collected by the operators may be recorded on check sheets during the round. Any deviations or discrepancies in the process are raised as the operator returns to the facility control central.

5.5.5.4 Condition Monitoring Surveillance Round

Condition monitoring surveillance rounds may be conducted by the FI&R group and are particularly focused on critical equipment. Usually these rounds are very detailed and could take up to a day for each round to be completed, depending on the amount of critical equipment and detail of each facility area. Typically, condition monitoring surveillance rounds are conducted on a monthly basis.

More often than not, the information is recorded directly onto a handheld device because of the volume of data that is collected and uploaded to a computer once the round is complete. The data is then analyzed and a report issued as necessary once the condition monitoring technician completes the round.

Safety & Reliable Operations	
Maintenance Rounds Checklist	
Zero Tolerance to Equipment Failure	*Report any equipment variances*
General Check for All Plant Items	
Gauges - Replace any damaged / inaccurate gauges	
Guards - Check seal and coupling guards are in place and secure	
Clean bedplate - Remove debris and oil accumulation	
Housekeeping - Check area for good housekeeping, tidy as required	
Labelling - Ensure correct labelling of all equipment	
Piping - Check piping for leaks, frayed, damaged or deteriorated hoses	
Lagging - Where applicable check all lagging for looseness in place and in good condition	
Safety valves and other safety relief devices are not obstructed	
Fasteners are tight, electrical leads secure and in good order	
Motor Checks	
Listen for abnormal noises	
Check for heavy vibration of adjacent structural components or driven machine	
Listen for rubbing of fan / debris caught in the fan cooler	
Feel for air flow over the motor finned surface or through duct	
Feel for air movement through the shroud. (Will also indicate correct rotation)	
Clean any debris and dirt away from shroud area	
Verify the shroud is secured to the motor	
Check motor bearing for high vibration & temperature	
Look for smoke from the motor / bearings	
Pump/Vacuum Pump Checks	
Check for visible leakage from casing	
Listen to bearings for high or abnormal noise	
Hand check pump / gearbox / turbine / motor bearings for high vibration & temperature	
Listen for cavitation, popping or grinding on pump casing	
Look for smoking from the case, bearing housing, motor & under insulation	
Look at discharge pressure for a sporadic or abnormal reading	
Check oilier bottle for level & quality	
Check for discolouration of material around the seal	
Vessel / Reactor / Dryer Checks	
External surfaces condition is free from corrosion and painted surface of good quality	
Prime flanges and stud bolts free from corrosion	
Prime Isolation valves are free from corrosion, leaks and have freedom of movement	
General look round vessels for leaks and abnormalities, look for excessive liquid, oil leaks	
General look round for obvious excessive movement or vibration of structural steel & piping	
Listen for abnormal noises (including gearbox)	
The barrier fluid is clean and free from debris	
Check drive belts for vibration (flopping) & squealing and no belts missing	
Centrifuge Checks	
Pipes, hoses, expansion joints and cylinders are free of corrosion, leaks and breaks	
Pressure and temperatures of hydraulic systems are within tolerances	
Check the condition of the contamination indicator of hydraulic system (Green = Good)	

Figure 5.7 *Maintenance rounds checklist.*

5.6 MAINTENANCE WORKFLOW

The maintenance planning and scheduling workflow process is illustrated in Figure 5.8. The workflow provides an overview of the principal steps involved in planning, scheduling and executing maintenance work orders in a facility.

Maintenance organizations often tend to have very detailed workflow processes to capture all eventualities in defining, prioritizing, planning and execution work, which build on the basic workflow presented in

Figure 5.8 *Planning and scheduling workflow.*

Figure 5.8. A more detailed and custom maintenance workflow is necessary in order to document how the maintenance organization carries out
its maintenance work. This also creates visibility in the information flow
into and out of the maintenance group into other groups such as operations and FI&R.

The basic maintenance planning and scheduling workflow is made up of the following principal steps, which are detailed in the following sections:
- Identify a need for work;
- Screen requests for work;
- Determine if a task plan is required;
- Develop a task plan;
- Procure materials;
- Schedule the maintenance work;
- Prepare the equipment for maintenance;
- Perform the work order;
- Update the equipment history;
- Complete and close the work order;
- Update the task plan records.

5.6.1 The Requirement for Maintenance Work

The requirement for maintenance work can materialize in a number of ways. It may simply be a case of facility employees of all disciplines identifying a fault on equipment items or piping, such as leaks, noisy equipment, damaged equipment, etc., during the normal course of their day. Work may also be required as a result of a condition monitoring system pointing to equipment deterioration. The management of change process may result in a requirement for maintenance work to include small modifications to facility equipment, for example through the engineering teams. Finally, much of the maintenance backlog is a result of preplanned work that is preprogrammed into the CMMS as part of a preventive maintenance program.

The nature of the maintenance work must be clearly communicated and documented, which can be achieved in a number of ways. It is very important and an integral feature of the FIEM to document work requests so we can keep track of equipment history. Work orders are generated automatically (preauthorized work requests) by the CMMS for most maintenance work, including preventive maintenance tasks and condition-monitoring tasks or routines. Caution needs to be taken when the need for reactive work is identified that requires an immediate response. In this situation, it is tempting to bypass the planning process with the intention of documenting it later. This action leaves us open to mistakes and potential risks and should be avoided.

Maintenance work may be identified and requested by the vast majority of personnel on the facility. It is done so by the requester raising a work request in the maintenance management system, usually the CMMS.

5.6.1.1 Maintenance Work Request

A work request is a notification to perform maintenance work in a facility. If approved, the work request will ultimately be converted into a work order and is used to record equipment and maintenance history.

There may be a number of different types of maintenance work requests, such as:

- Routine Maintenance Request;
- Maintenance Repair Report;
- Minor Maintenance Work Request;
- Test and Inspection Request.

5.6.1.2 Maintenance Work Order

A maintenance work order is an approved work request for maintenance work to be performed at a facility. The maintenance work order comprises the specific details for the maintenance work to be performed.

There may be a number of different types of maintenance work orders, such as:

- Normal Maintenance Work;
- Preventive Maintenance Work;
- Predictive Maintenance Work;
- Test and Inspection Maintenance Work;
- Minor Maintenance Work;
- Management of Change Maintenance Work.

5.6.2 Screen Requests for Work

All maintenance work requests are screened. This is to ensure that resources are expended wisely and there is value in performing the maintenance work. The screening process may result in a number of outcomes, including an approval to proceed with the work, rejection of the request due to a number of reasons, and a request for additional information to support the work request.

Maintenance work requests should be screened by competent facility personnel who are responsible for facility maintenance costs and facility production. This is because the work that is approved, deferred or rejected will have an impact on the overall cost and resource utilization at the facility. The number of maintenance work requests that are approved represents the maintenance backlog for the facility.

5.6.3 Determine if a Maintenance Task Plan is Required

It is important that all maintenance work is planned. For very basic types of maintenance tasks the level of planning may be much less. In some cases, it may even be appropriate for planning to be performed by the maintenance technician, which eliminates the additional work for a trained and experienced maintenance planner. More complicated maintenance tasks certainly need to be planned by a planner. In these cases, there are usually a number of factors involved, such as special tooling or equipment like heavy lift cranes to be scheduled, engineering input such as drawings or calculations, coordination for simultaneous work between multiple work groups, etc.

5.6.4 Develop a Maintenance Task Plan

The maintenance task plan is normally prepared by a maintenance planner. It is important to ensure that there is a sufficient amount of detail in the plan so that the work scope and method statement are clear. Safety precautions are factored into the plan along with the tools and materials required. An example of a maintenance task plan is shown in Figure 5.9. The task plan generally covers the following information:

- Work order summary details;
- Safety and permit requirements;
- Description of the maintenance work to be performed;
- Special personal protective equipment (PPE) needed;
- Materials and equipment plan.

One of the key principles of the FIEM is that there is a feedback loop on the key workflow processes. Since maintenance task planning is a key process, it is necessary to ensure that this process is executed well and audited for continuous improvement. Therefore, a planning package check is required to ensure that the work order plan along with the planning and maintenance personnel involved in the work order planning process meet the required quality mark. An example of a planning package development checklist is shown in Figure 5.10.

5.6.5 Procure Maintenance Materials

One of the integral enablers for a smoothly functioning maintenance management system is an efficient and effective procurement department for maintenance spares and materials. The procurement department should ensure that there is a complete and current database of spare parts for facility equipment as defined by the maintenance strategy. The spare parts

Maintenance Task Plan					
Work order no.:	**Work order description:**				
Requested by:	**Requested date :**		**Required date:**		
Priority:	**Planner name / contact details:**				
Activity type:	**Location:**				
Work description:					
Safety requirements			**Procedures required**	**Permits required**	
1 Hot work					
2 Confined space					
3 Toxic gas zone					
4 Radiation					
5 Other					
Task plan					
Activity description			**Craft**	**No. Craft**	**Duration (Hours)**	**Total (Hours)**
1					
2					
3					
4					
5					
Special PPE and tools					
1					
2					
3					
4					
5					
Material plan					
Item description				**Quantity (No.)**	
1					
2					
3					
4					
5					

Figure 5.9 *Maintenance task plan.*

and materials must be stored in a secure area that factors safeguards from weather and adverse environmental conditions.

The procurement of maintenance spare parts and materials can have a sizeable impact on the maintenance operating budget. It is often cost effective to consider prearrangements with key suppliers for spare parts or materials for expensive items or high throughput consumable items.

5.6.6 Schedule the Maintenance Work

The effective scheduling of maintenance work is important to ensure that the maintenance team is fully utilized, important activities are prioritized, and that maintenance work is well organized and communicated to the relevant parties within the facility organization.

Scheduling of maintenance work is often done on both a weekly and a daily basis, with scheduling meetings held with key stakeholders representing

No.	Planning Package Development Checklist	Check
1	**Work Order**	
	Work order number is assigned	
	Work has been appropriately prioritised	
	Work schedule date is defined	
	Work scope has been signed off	
2	**Work Order Review**	
	Field assessment of the work is complete	
	Work plan has been prepared for maintenance work	
	Work plan is approved by the maintenance supervisor	
	Material requirements are defined and available	
	Safety precautions are in place	
	Tools are defined and available	
	Scaffold has been installed and registered (as required)	
3	**Work Order Documentation**	
	Documentation is complete and available at the work site	
	Engineering standards have been adhered to	
4	**Special Conditions**	
	Management of change proposal is approved	
	Live electrical work plan has been approved	
	Hydro test work plan has been approved	
	Critical lift work plan has been approved	
	Confined space work plan has been approved	
	Welding on live equipment work plan has been approved	
	Excavation work plan has been approved	
	Equipment vendors are available at the work site (as required)	
5	**Operations Preparation**	
	Equipment isolations are complete	
	Blinding of process equipment is complete	
	Purging of process equipment is complete	
6	**Equipment Date**	
	Equipment manuals are available at the work site	
	Equipment and material weights are identified	
7	**QA/QC Requirements**	
	Inspector is identified and aligned	
	Welding (heat treatment) details are identified	
	Welding (radiography) details are identified	
	Hydro test details are identified	
	Painting requirements are identified	

Figure 5.10 *Planning package checklist.*

the various groups in the facility. These attendees would normally include: maintenance manager; planner; maintenance work requesters; representatives from the operations group for facility production planning, and if there are test and inspection tasks, representatives from FI&R.

The main aim of the weekly scheduling meeting is to coordinate maintenance tasks with the production plan from the operations team and to set priorities for work to be executed during the following week. The daily meeting is held to review maintenance task progress and coordinate with the operations preparatory work in order to ready the facility for the maintenance teams. During the daily meeting, work is reviewed and set for the next 24 hours.

5.6.6.1 Work Order Priorities

Maintenance work order priorities are usually set based on the criticality of the work to be done and the facility production schedule. It is important that work order priorities are reviewed and agreed between operations, maintenance and the FI&R groups so that a collective viewpoint can be considered and all functional groups have awareness of what work is being done on the facility. Work order priorities also drive the appropriate level of planning and scheduling required. Critical work may require a more detailed level of planning, for example. It is important that maintenance work is schedule in accordance with the agreed predetermined priorities. It is also important to be aware that the highest priorities, which usually represent unplanned tasks and are also usually the higher cost activities, are minimized. This is so that the focus and effort is expended on the planned work. We shall review the mix of planned and unplanned work in section 5.8.

There are a number of ways to prioritize maintenance work orders. Maintenance work order priorities can be categorized as follows.

Priority 1 Work Orders

Priority 1 work orders are regarded as emergency maintenance tasks which should be executed without delay and continued through to completion. Overtime usually has a blanket approval to proceed for Priority 1 work orders.

Priority 1 work orders are usually in response to the following circumstances:

- Extremely high potential for health hazards, safety risk or a major environmental release;
- Facility products that cannot meet existing sales quality specification or capacity;
- Cannot meet existing laws or regulations.

Priority 2 Work Orders

Priority 2 work orders can be defined as urgent work which should be completed as soon as possible, typically within 48 hours. Overtime approval is usually not required for priority 2 work orders.

Priority 2 work orders are usually in response to circumstances where:

- There is a high potential for health hazards, safety risk or a major environmental release;
- There is a high potential that the produce quality will not meet the sales specification;
- There is a high potential for a critical equipment failure.

Priority 3 Work Orders

Priority 3 work orders are usually defined as essential work that can be re-scheduled. Priority 3 work orders should be scheduled and executed within 1 week of the work order request.

Priority 3 work orders are usually in response to circumstances where:
- There is a potential for personnel injury that needs to be corrected;
- A preventive maintenance task: inspection or testing;
- There is a potential product quality impact that may escalate.

Priority 4 Work Orders

Priority 4 work orders are usually defined as work that should be performed during a scheduled shutdown period. There is no urgency to execute priority 4 work orders which may be delayed for a number of months. They may require design work, which may require coordination with the operations team and potentially equipment vendors.

5.6.6.2 The Scheduling Meeting

The scheduling meeting is a core element of the maintenance planning and scheduling workflow. Typically, the weekly scheduling meeting will last approximately 30 minutes and a daily scheduling meeting of approximately 15 minutes.

A representative agenda for a weekly or daily scheduling meeting is as follows:
- Review of new maintenance work added;
- Review priority 1 maintenance work orders;
- Review and reschedule of maintenance work not completed;
- Assess the reasons for not completing the scheduled maintenance work and record the lessons learned;
- Review the production requirements for this weeks' schedule;
- Review of maintenance work available to schedule;
- Review of old maintenance work for possible cancellation;
- Review and address maintenance manpower and materials constraints.

5.6.7 Prepare Equipment for Maintenance Work

There are usually a number of preparatory activities that are required prior to commencing with maintenance work. These may include: preparation of work permits for the maintenance work, shutdown of the equipment, making the equipment safe for maintenance (isolation or lock out of moving parts, etc.). This work is usually performed by the operations team and efforts are made to ensure the preparatory work is completed ready for the next maintenance shift in order to avoid any lost time.

5.6.8 Perform the Maintenance Work

Once the operations team has completed their preparations for the equipment the maintenance team can then perform the work order. The work should be managed by a competent maintenance team leader and performed by a team that is made up of the necessary disciplines.

Throughout the duration of the task, the maintenance team should keep the planner and operations team updated on the progress and highlight any potential delays due to unanticipated events. The team should also ensure the maintenance work is updated along the way, in case there are any points of interest for the FI&R teams.

5.6.9 Update the Equipment History

One of the key features of the FIEM is the need to maintain an up-to-date facility equipment database, which includes a comprehensive and robust set of records for facility equipment performance.

Once the maintenance work is completed for each work order, it is imperative that the work performed on an item of equipment and any points of interest are recorded in the equipment history database. This information may be critical and may be used by the facility teams for future troubleshooting and reliability improvements. The equipment history records should cover at a minimum the following information: current equipment condition; maintenance work performed; spare parts used and any points of interest (obvious wear or degradation mechanisms, markings, decolorization, etc.). If the maintenance work is in response to equipment failure, a failure investigation must be carried out. The failure investigation matrix shown in Figure 4.14 can be used to determine the type of failure investigation and resource expenditure that may be required.

It is an effective control to ensure a senior maintenance engineer signs off on the work orders to ensure the equipment history is updated to a sufficient degree of detail. Furthermore we can develop an audit as a feedback mechanism to ensure this system is robust. We shall discuss audit and review in Chapter 7, section 7.3, "Quality Assurance."

5.6.10 Complete and Close the Work Order

After the work is performed and the equipment history is updated, the maintenance work order can be closed. In doing the work originator (work requester) should be notified that the task has been completed and any findings communicated. It is then important that the work requester reviews the work completed (at the site of the work) along with the maintenance team leader to ensure the work has been completed as required and it meets the quality standard. If this is the case, the work order must be closed and no further cost allocated.

5.6.11 Update the Task Plan Records

The nature of the majority of maintenance work is that it is repetitive. For example, equipment overhauls, spare parts replacements, filter and oil changes, etc. are completed periodically through either condition-based or time-based maintenance. These task plans are in many cases repetitive. In this case, in the interest of improving facility maintenance efficiency and effectiveness, a library of task plans can be created. This library can be used for planners to draw relevant task plans for the maintenance team. An example of a maintenance task plan is shown in Figure 5.9.

As part of the FIEM continuous improvement efforts, the task plan must be revisited after each maintenance task is closed. The task plan review and update should cover any improvements to the way the task is performed in terms of efficiency and safety, as well as any lessons learned such as unforeseen events that may have caused delays and that may occur again.

5.6.12 Guidelines for Planning and Scheduling System

Maintenance planning and scheduling is paramount to an effective and efficient maintenance management workflow and FIEM. If implemented well, the planning and scheduling workflow can substantially reduce facility equipment downtime. It will ensure that preventive maintenance is completed on time and there will be a noticeable reduction in equipment failures. In addition the workflow can make step change improvements to the maintenance organization in terms of reducing the number of maintenance team members required to perform the maintenance work and improving the efficiency of the work performed.

In order to ensure the planning and schedule workflow is effective, there are a number of activities that should be completed in order to ensure it is appropriately supported. First, a planning and scheduling manual that accompanies the workflow and clearly identifies the associated activities, including roles and responsibilities, should be developed. This manual will ensure expectations are clear for all of the stakeholders. It is also important that all of the stakeholders are sufficiently trained on the process and systems.

In the interest of the PDCA cycle and the FIEM continuous improvement, a thorough planning and scheduling audit and review schedule for the different steps in the process must be developed. This is so that we can learn from our current performance and address any inefficiencies or errors on an ongoing basis.

5.7 MAINTENANCE STRATEGY

Selecting a successful maintenance strategy requires a good knowledge of maintenance management principles and practices as well as knowledge of specific facility performance. There is no one correct formula for maintenance strategy selection and, more often than not, the selection process involves a mix of different maintenance strategies to suit the specific facility performance and conditions.

There are a number of maintenance strategies available today that have been tried and tested throughout the years. These strategies range from optimization of existing maintenance routines to eliminating the root causes of failures altogether, to minimize maintenance requirements. Ultimately, the focus should be on improving equipment reliability while reducing cost of ownership.

An effective maintenance strategy is concerned with maximizing equipment uptime and facility performance while balancing the associated resources expended and ultimately the cost. We need to ensure that we are getting sufficient return on our investment.

Are we satisfied with the maintenance cost expended versus equipment performance and uptime? There is a balance to be had in terms of maintenance cost and facility performance. We can develop a suitable maintenance strategy to help tailor this balancing act in order to ensure the return on investment is acceptable (Figure 5.11).

A maintenance strategy should be tailored specifically to meet the individual needs of a facility. The strategy is effectively dynamic and must be updated periodically as circumstances change. The strategy must include a detailed assessment of the current situation at the facility and consider the following questions:

- What is the performance history of the facility equipment and systems?
- What are the production targets, i.e. what are the mission times for facility equipment and systems?

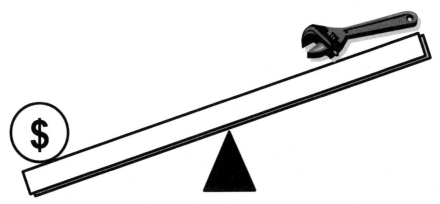

Figure 5.11 *The maintenance balancing act.*

- What are the facility shutdown targets?
- What is the current maintenance budget?

Once we have clarity on the current situation and constraints, we need to define the objectives of the maintenance plan. The objectives must align with the business objectives of the company. They must be developed by all of the key facility stakeholders and be clear, concise and realistic. There may be a number of components to the strategy objectives – for example: improve equipment uptime, reduce maintenance costs, reduce equipment operating costs, extend equipment life, reduce spare parts inventory, improve MTTR, etc

An example of a maintenance strategy workflow is illustrated in Figure 5.12. This workflow is developed to optimize and improve an existing facility maintenance program. Depending on the specific circumstances at the facility, our strategy may also take us into the direction of a step change approach to maintenance management and opt for a reliability-centered maintenance (RCM) program, which may replace our existing maintenance program. This strategy is labor and time intensive and can be expensive; we will discuss RCM in section 5.7.1.

It is a common theme in the industry that maintenance budget and resources are very thin on the ground relative to the amount of work that needs to be done. Therefore, prioritization of maintenance resources is absolutely essential in order to be successful. Once we have defined our maintenance strategy objectives, we need to define facility equipment criticality. We have discussed the concept of criticality in Chapter 4. Criticality is a risk-based approach that can help us to prioritize our resources effectively. It can also help to appraise the requirement and effectiveness of maintenance tasks already populated in the MMS or CMMS.

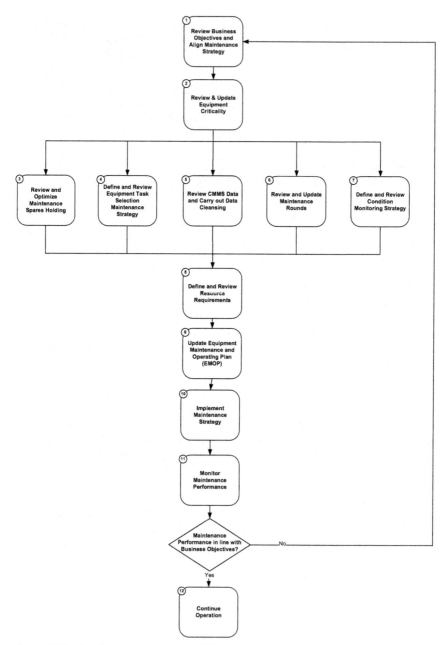

Figure 5.12 *Maintenance strategy workflow.*

The output of the criticality review provides an input into a number of maintenance strategy activities, as shown in Figure 5.12.

These activities may include the following:

- Reviewing and optimizing maintenance spares holding;
- Defining and reviewing equipment tasks selection, such as corrective, proactive and run to failure;
- Review and update maintenance rounds;
- Define and review the condition monitoring strategy;
- Review the CMMS to assess the effectiveness of maintenance tasks (data cleansing).

The key principle behind the review and optimization of an existing maintenance strategy is an accurate and robust criticality assessment, which may impact many of the strategy objectives.

Another common theme in the industry is that many computer maintenance management systems are populated with a large proportion of preventive maintenance tasks that may be considered as superfluous and even not necessary. These tasks may consume a large proportion of the maintenance resources and time without an acceptable return on the investment made (maintenance cost). The maintenance strategy should also ensure the current data in the CMMS is value adding and therefore carry out a "cleansing" exercise. A data cleansing exercise critically reviews and appraises the current CMMS tasks and aims to eliminate the tasks that may not be adding value and therefore are superfluous. By focusing on equipment criticality, these activities can be reviewed and appraised in a logical and systematic way.

Once the equipment criticality assessment is completed and the strategy objectives have been reviewed and updated, maintenance resources can then be aligned to the strategy. The maintenance strategy objectives will dictate the resources and associated maintenance costs. The next step in the strategy development process is to update the equipment maintenance and operating plan as presented in section 5.5. The EMOP is the primary record and source of maintenance and operation information of each equipment item and includes the up-to-date maintenance and operating strategies. It provides the baseline information including equipment maintenance and operating parameters. We are then in a position to implement the maintenance strategy on the facility.

It is important to understand the impact (and the success) of the new maintenance strategy. This is achieved by setting key performance indicators (KPIs) to assess the facility maintenance performance. This is done by first developing a benchmark data set. How is the facility currently performing? What is the cost of maintenance? What is the MTBF? What is the MTTR? What is the maintenance rework ratio? Once the current facility maintenance performance is benchmarked, we can then measure maintenance performance against this benchmark. Maintenance performance is reviewed periodically and, depending on the results, may be reviewed and updated more frequently. We will look at facility integrity KPIs and dashboards in Chapter 9.

If the maintenance performance is in line with business objectives, then the facility operation will continue; however, if there is any deviation in performance or change in the facility process or criticality ranking, then the maintenance strategy should be revisited.

5.7.1 Reliability-Centered Maintenance

In 1978 Stanley Nowlan and Howard F. Heap published a report aimed at determining new and more cost-effective ways of maintaining complex systems in the aviation industry. It was called "Reliability-Centered Maintenance" (RCM) [5.2].

Today, reliability-centered maintenance (RCM) is used across many industries and is recognized as one of the leading practices for oil and gas and petrochemical facility maintenance. RCM acknowledges that all equipment in a facility does not have an equal importance and that there are significant advantages in prioritizing maintenance efforts on certain facility equipment. RCM effectively provides a structured approach to the development of a maintenance program. It focuses on equipment needs and ultimately results in a well-grounded basis for facility maintenance with a high proportion of proactive maintenance. RCM addresses the basic causes of equipment and system failures. It aims to ensure that controls are in place to predict, prevent or mitigate these functional failures and hence the associated business impact [5.3]. RCM is defined by a technical standard from the Society of Automotive and Aerospace Engineers (SAE), namely SAE JA1011 (1999) [5.4].

5.7.1.1 RCM Workflow

Reliability-centered maintenance (RCM) analysis provides a structured framework for analyzing the functions and potential failures of facility equipment, such as pumps, compressors, a facility processing unit, etc. The emphasis of the analysis is to preserve system function, instead of focusing on preserving the actual equipment. The output of an RCM program is a series of scheduled maintenance plans. The RCM standard, SAE JA1011, describes the minimum criteria that a process must comply with to qualify as an RCM Process [5.4].

Although in the application of RCM there tends to be a large amount of adaptation, usually it follows the steps illustrated in the workflow in Figure 5.13.

Make Preparations for the RCM Analysis

In order to ensure the RCM analysis is executed smoothly, there are a number of preparatory activities that should be completed in advance of an RCM analysis.

First the RCM team should be carefully assembled. The team should comprise a cross-section of facility operations, maintenance and FI&R teams with a strong technical understanding of the equipment to be analyzed. The team should also be conversant with the RCM analysis methodology.

RCM analysis requires a large investment of time and resources. Given this, it is often necessary for the facility maintenance group conducting the analysis to focus on a selection of equipment or systems. The equipment or systems to be analyzed should be identified and boundaries drawn around the battery limits of systems. This is to ensure clear demarcation of the RCM scope so that efforts and time are directed appropriately. It is often the case that a criticality assessment is used to determine the equipment or systems selection.

Determine the Functions and Potential Functional Failures

Reliability-centered maintenance focuses on preserving equipment functionality. The next step in the process is to determine the function or functions that the equipment or systems are intended to perform. Equipment functions should also be prescriptive in the definition of a function and include performance limits, for example.

Once the functions are clearly defined by the RCM team, their corresponding potential functional failures are defined. Functional failures may also include poor performance of a function or overperformance of a function as well.

Identify and Evaluate the Effects of Failure

The next step in the process is to identify and evaluate the effects of the equipment failure. This step enables the RCM team to prioritize and choose an appropriate maintenance strategy that can tackle the failure. It is

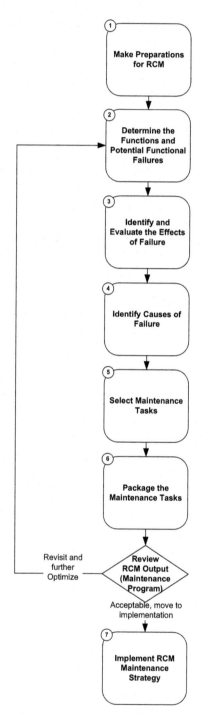

Figure 5.13 *Reliability-centered maintenance workflow.*

common to employ a logic diagram to structure this part of the process in order to consistently evaluate and categorize the effects of failure.

Identify Causes of Failure

By identification of the specific cause of the failure we are able to understand the root cause and ultimately define an appropriate maintenance strategy that can address the failure altogether.

It is important to leverage the skills and experience of the RCM team in order to ensure the cause of the failure is clear and accurate. The cause of the failure should be described in sufficient detail at this stage. This is so that we are able to ensure the maintenance task selection step in the process is confidently and reliably completed. It may be appropriate to refer to the RCM standard, SAE JA1012, which presents useful guidance as to how to identify causes of failure [5.4].

Select Maintenance Tasks

At this stage in the process, we have identified the functions that equipment is intended to perform and the ways that these functions could fail. We have evaluated the effects of functional failures and identified their causes; the next step in the RCM process is to select appropriate maintenance tasks for the equipment to prevent such failures. There are a number of ways to carry out this exercise; however, the RCM team's skill set and knowledge is the key factor.

Package the Maintenance Tasks

The final step in the RCM process is to package the maintenance tasks into a practical and robust maintenance program. This process involves reviewing the selected maintenance tasks and grouping them in a logical way so that they can be uploaded into the facility CMMS. The ultimate goal in packaging the RCM tasks is to arrive at a practical and efficient maintenance program.

5.7.1.2 *Implementation of RCM Maintenance Strategy*

Reliability-centered maintenance (RCM) has been in use for a number of years. It provides a structured and systematic framework which can result in an effective maintenance management program for facility equipment.

It is no surprise that RCM is a resource intensive and time-consuming process that can be expensive to develop and implement. There are a number of iterations of RCM that attempt to reduce the effort needed to develop and implement an RCM program, with varying degrees of success. It is important to maintain the key principles of RCM and not to overstretch the battery limits that were agreed on by the facility maintenance team at the start of the process. This may lead to disillusionment and frustration and eventually may result in a failed implementation effort.

The approach to the development and implementation of an RCM maintenance program must be executed with dedication and tenacity. It is also important for the facility management team and the wider facility functional groups to buy in and support the RCM implementation effort.

5.7.2 Failure Mode and Effects Analysis

Failure mode and effects analysis (FMEA) is a useful and practical tool for analysis of equipment failures. FMEA, which dates back to the 1940s, was one of the first techniques used as a methodical approach to failure analysis.

It was initially developed by the US military to address problems with the premature failure of military equipment and systems. It is detailed in MIL-P-1629, which is a US Armed Forces Military Procedure [5.5]. FMEA has evolved over the years and is now extensively used across a number of industries including space agencies, food service, software, healthcare, petrochemical and oil and gas. FMEA may also be referred to in standard SAE J1739 (Potential Failure Mode and Effects Analysis in Design) [5.6] and standard IEC 60812 (International Standard on Fault Mode and Effects Analysis) [5.7]

FMEA can also form part of a reliability program such as an RCM study. It involves reviewing equipment systems, subsystems and components to identify failure modes, their causes and their effects. The effects analysis involves examining the consequences of the failures on the particular equipment systems, subsystems or components. For each subsystem or component the failure modes and their resulting effects are recorded in an FMEA worksheet. There are many variations of FMEA worksheets to record the output of the analysis.

The main steps involved in the execution of an FMEA are as follows:
1. Define the objectives and the expectations of the study;
2. Identify and ensure the battery limits are clear of the equipment or system that is to be analyzed;
3. Define the equipment systems, subsystems and components and their relationships;
4. For each equipment system, subsystem or component, identify the failure modes, their causes and effects.

FMEA studies are particularly useful when applied to specific equipment or systems. This is because the tool is designed originally for standalone military equipment. We may also wish to target the FMEA study on specific equipment or systems may have been highlighted as the result of a criticality analysis.

5.7.3 Planned Maintenance Optimization

Planned maintenance optimization (PMO) is a well-established, tried and tested maintenance strategy, dating back to the 1990s. Around this time there was a lot of concern from the industry that RCM did not suit the requirements for facilities that had existing maintenance programs with limited resources and timescales to perform an RCM study. This is because primarily RCM is a tool that is designed for use in the design stage of the facility life cycle. PMO, on the other hand, is specifically designed to target existing maintenance programs.

The PMO process is illustrated in the workflow in Figure 5.14. PMO identifies planned maintenance database activities from an existing facility CMMS, categorizing them into planned maintenance craft groups. The workflow then reviews each corresponding facility equipment history to determine if the planned maintenance task is necessary. These tasks are critically evaluated and ultimately optimized based on the added value. Finally the maintenance program is updated along with the CMMS.

A PMO study may be conducted manually in a task force team or by employing commercially available software. There are numerous PMO software titles available in the market, some of which can be interfaced with a CMMS. Typically the decision to implement a PMO strategy is made in an ad hoc fashion by the maintenance management team. It is usually driven by budget and resource constraints.

5.7.4 Defect Elimination

"Failures result from defects. Eliminating defects is the way we improve constantly and forever the system of production and service" [5.8].

Equipment failures are a result of defects; therefore by eliminating defects we can improve equipment reliability. Defect elimination is a maintenance strategy that takes us back to design. It aims to prevent defects being introduced at the early stages of the equipment life cycle, thereby removing the defects during the operational stage of the equipment life cycle.

By eliminating the defects that have potential to cause future equipment failures, maintenance requirements will also be reduced, resulting in improved equipment uptime. Defect elimination can actually reduce the maintenance requirements on equipment or systems and hence lower maintenance cost.

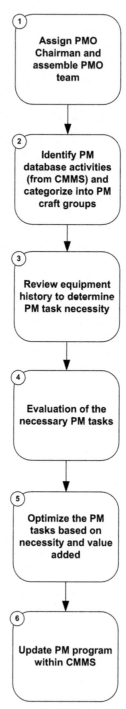

Figure 5.14 *Planned maintenance optimization workflow.*

Defect elimination aims to identify failure modes and eliminate them at the outset. Each part of the equipment is taken in its component parts and corresponding defects are identified. Mitigation plans are then prepared for each and every defect identified in order to eliminate the failure mode. Control measures and quality assurance standards are developed in order to detect and eliminate defects before they are designed into the equipment and systems. One of the methods that could be employed in defect elimination is the FMEA tool as presented in section 5.7.2, which is based on failure mode and effects analysis.

Preemptive maintenance strategies such as defect elimination are very useful because they can be very cost effective and have lasting impressions on reducing maintenance requirements.

5.7.5 Intentional Overdesign Selection

In some instances maintenance managers may decide to purposefully overdesign a particular equipment or system on the facility. The idea behind this is that these particular equipment items or systems are therefore able to withstand deterioration processes more and function for longer periods of time between failures.

This decision may be made when dealing with highly critical processes on the facility, such as processing toxic or hazardous materials or chemicals, or where there is a requirement to increase the reliability of a certain part of a process that may warrant additional robustness of equipment design.

This is a strategic maintenance decision intended to prolong facility equipment and systems life and therefore maintain longer periods of production. It involves increasing the design specification of equipment or systems with more robust parts, higher specification materials of construction, better surface protection coatings, etc.

Maintenance management is a continuous improvement process. The intention is to add value by improving equipment reliability while reducing cost of ownership. Clearly there is a balance to be had with cost of ownership versus additional value added, particularly with this strategy. There may be a higher cost of ownership; however, this is offset against the improvements to the production output.

5.7.6 Shutdown Overhaul Maintenance

During shutdown overhaul maintenance equipment and systems are repaired or overhauled on a set frequency that is shorter than the MTBF. By doing this we should prevent an unexpected failure.

Such work is typically done as an overhaul, where the whole of the equipment is removed from operation during a shutdown and taken to the workshop to be stripped down to its component parts and rebuilt as new.

Use of shutdown overhaul maintenance strategy is aimed at ensuring uninterrupted production for a specific period of time. By renewing or overhauling equipment regularly we remove the wear-out related stoppages. Once equipment is overhauled to manufacturer's standards we can expect as-new performance. However, we are also exposed to "infant mortality" risks due to poor quality control, mistakes during assembly, incorrect material selection and introduced damage.

5.8 THE RIGHT MIX

There is no perfect maintenance strategy to suit all facilities and all circumstances. More often than not, a number of different maintenance strategies are blended to meet the specific needs of the facility business objectives. This is usually dictated by a number of reasons, tailored to the individual needs of the facility.

There will most probably be a need for corrective maintenance to deal with equipment failures and corresponding root cause analyses. There will also likely be a requirement for predictive maintenance to deal with aging equipment along with preventive maintenance for routine maintenance tasks.

We have explored options for maintenance strategy development in Figure 5.12 and noted that one of the key elements of maintenance strategy development is the requirement that strategy must evolve as circumstances change and therefore the blend of maintenance strategies must also change.

There is no specific guideline or standard as to what the "right mix" of maintenance strategies should be. Each and every oil and gas and petrochemical facility is different and therefore their corresponding maintenance strategy is unique.

There tends to be a general consensus in the industry that there should be a certain blend of preventive, predictive and corrective maintenance in order to have a successful maintenance operation, be efficient and also cost effective. Figure 5.15 shows the perceived "world class" maintenance strategy mix, which comprises a vast majority percentage of proactive maintenance and only a minority percentage of corrective maintenance.

On the other hand, there is a perception that the industry average maintenance facility strategy mix (Figure 5.16) consists of a vast majority of

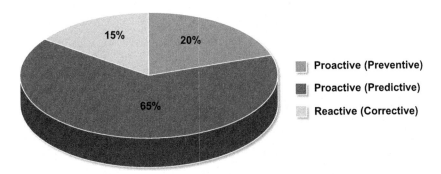

Figure 5.15 *"World class" maintenance strategy mix.*

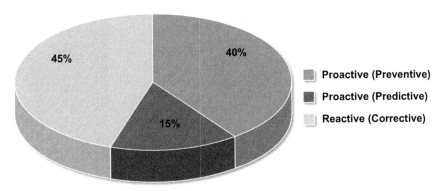

Figure 5.16 *Industry average maintenance strategy mix.*

corrective maintenance. This is an environment that we have discussed, "the reactive zone," where repairs are a stressed environment in response to random equipment failures.

There is also a large percentage of preventive maintenance tasks, which focuses on the time-based equipment maintenance at preset intervals.

Facility operations that have gotten their maintenance strategy mix right tend to have a cost-effective maintenance operation with high performing and reliable equipment.

CHAPTER 6

Operations

Contents

6.1 INTRODUCTION

The operations element of the facility makes up the final part of the facility segment of the FIEM (Figure 6.1). The integrity of facility operations is established by setting operating envelopes for the facility equipment and systems and ensuring the facility operates within these envelopes. This requires an effective facility operations plan to be in place for the various operating conditions for both normal and abnormal operating conditions. The facility operations plan is aligned with the overall business objectives, which may include production targets and major shutdowns.

There are a number of controls that must be in place to ensure the facility operations plan is adhered to. These include a detailed set of individual equipment maintenance and operations plans (EMOPs), operations procedures and practices, well-structured FI&R and maintenance programs, reliable facility critical equipment and a qualified and experienced operations team.

The operations team is tasked with ensuring that the facility operations plans and the individual EMOPs are executed satisfactorily via operating procedures and practices. The operating procedures should explicitly

Facility Integrity Management
http://dx.doi.org/10.1016/B978-0-12-801764-7.00006-1

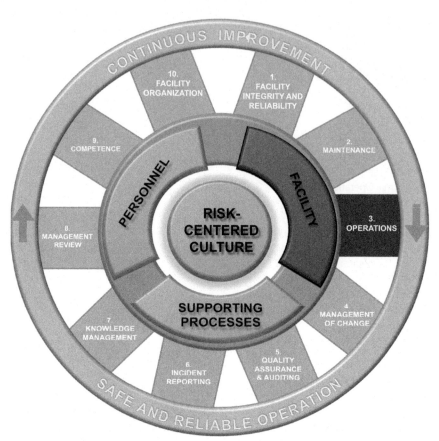

Figure 6.1 *The facility: operations.*

describe and document the facility operations processes for all equipment and systems at the facility.

Operating procedures should detail the process activities required for each and every eventuality of the facility, from start-up of the facility, to normal operation to normal shutdown. The procedures should also include process activities in the event of an emergency, such as emergency operating procedures and emergency shutdowns. Finally, there may be circumstances whereby the facility is required to operate under unusual conditions and therefore temporary operating procedures must be developed, for example changes made during capital projects or commissioning work, etc.

Operating procedures must also consider health, safety and environmental (HSE) in all activities. A thorough health, safety and environmental control plan should be developed to address topics such as properties and hazards associated with the facility process fluids and chemicals. Corresponding precautionary measures should also be developed in the event of loss of containment of hazardous facility process fluids and chemicals, for example operational controls, personal protective equipment, etc.

6.2 FACILITY START-UP

Facility start-up is an important activity in facility operations. There may be a number of circumstances as to why the facility was shut down in the first place and the corresponding start-up procedure should consider these circumstances. This may be as a result of the initial start-up of the Facility after it was first commissioned, or due to a process feedstock change-out such as catalyst or low demand. More often than not, the shutdown may have been as a result of maintenance work or capital project work requirements. This may have resulted from a planned or unplanned shutdown. In this circumstance it is imperative to ensure that the construction and commissioning punch list is cleared and signed off. The operations team should then carry out their round checklist such as check valve direction, instrument transmitter connections, etc.

The objective of the facility start-up process is to safely and efficiently achieve a steady-state operation. There may be an extended commissioning exercise required to optimize the facility process to arrive at a steady-state operation. At this point the operating procedures may be modified under the management of a change process to record these optimum operating parameters.

6.2.1 Pre-Start-Up Safety Review

Recognizing that there are severe consequences as a result of a disorganized and poorly executed facility start-up, a pre-start-up safety review (PSSR) is common for oil and gas and petrochemical facilties. The PSSR provides a final check of the facility, particularly if there have been any new or modified equipment or systems introduced to the facility. The PSSR provides a systematic check for all process safety and equipment operating parameters. It ultimately aims to ensure the facility is ready and safe to start up, ensuring that any changes have been appropriately carried out; this includes:

- Actions and recommendations from safety studies such as hazard and operability studies (HAZOPs) are implemented;
- Modifications meet the required standards and specifications;
- Equipment consumables are in place such as oil, hydraulic fluids, etc.;

- Procedures have been updated and implemented, including emergency procedures;
- Training for new processes has been adequately provided;
- Equipment Maintenance and Operating Plans (EMOPs) are updated and have been implemented.

It is important that the PSSR is conducted by a multidiscipline team with representation from the operations, FI&R and maintenance teams in order to ensure the different skillsets and experience are considered during the start-up review. The output of the PSSR should be documented and filed appropriately so that reference can be made to it by the facility management team.

6.3 NORMAL OPERATIONS

During normal operation the facility operates as designed with the operating teams primarily tasked with ensuring that all equipment and systems are performing within their operating envelopes. Figure 6.2 illustrates a simple workflow process that describes the integrity focus for normal operations on a facility. In the event that the facility equipment or systems breach their operating envelopes, there is a higher risk of equipment failure, material wear and degradation and ultimately loss of containment. As the operations team identifies equipment or processes operating outside of their operating envelope, action needs to be taken.

The first port of call is to assess the risks associated with operating equipment or systems outside of their operating envelope. A decision must then be taken as to whether the risks can be managed. If not, a shutdown of the equipment may be required. Alternatively, the equipment operating envelope may be reviewed and redefined to accommodate the new operating envelope. This may be done on a temporary basis, for example in the event of certain circumstances such as commissioning equipment or even a permanent basis. In either case, changes in the operating envelope should be addressed through a management of the change process. We shall visit the integrity supporting processes and management of change in chapter 7.

6.4 ABNORMAL OPERATION

There may be situations at the facility whereby the facility process does not behave in "normal mode." In this situation there may be fluctuations in process parameters and conditions. The operation is monitored closely and abnormalities are evaluated in an attempt to arrive at a steady-state operation.

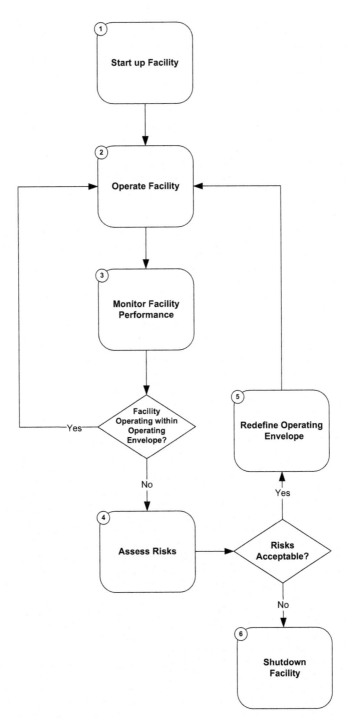

Figure 6.2 *Normal operations.*

6.5 OPERATING ENVELOPES

The central concept behind operating envelopes (OEs) is that facility equipment and systems are designed to operate within a specified range of process parameters. These parameters can be both physical and process related. If facility equipment and systems operate within these ranges (and are maintained in line with the maintenance program) it is probable that they will have a trouble-free operating life.

If there are deviations from these specified operating ranges, the probability swings in the opposite direction with the facility equipment or systems integrity compromised. Under these conditions it is likely that the facility equipment or systems will fail prematurely. The definition given to this concept is that of an equipment operating envelope (OE). OE specifies the process conditions within which the equipment will retain its integrity. All equipment or systems in the facility have an operating envelope that varies from process to process (Figure 6.3).

It is the role of the operations team to monitor and evaluate the facility operating envelopes carefully and continually. Deviations or *variances* must be identified as soon as possible, logged and communicated to the facility FI&R and maintenance teams and closely monitored.

6.6 EQUIPMENT MAINTENANCE AND OPERATING PLAN

We have discussed equipment maintenance and operating plans (EMOPs) in Chapter 5 with a maintenance bias. From an operations perspective, EMOPs are also the primary process as part of the FIEM defining operating

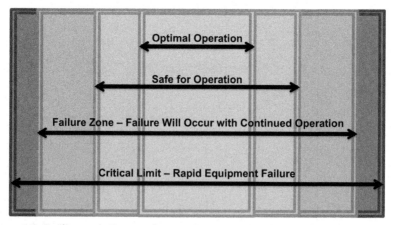

Figure 6.3 *Facility operating envelope.*

strategies and plans. EMOPs are designed specifically to bring together and maximize the synergies of FI&R, operations and maintenance, promoting values of the risk-centered culture along the way. EMOPs provide baseline information for all equipment operating integrity envelopes, as shown in Figure 5.6.

6.7 FACILITY MONITORING PRINCIPLES

In order to have a reliable facility with a world-leading uptime performance, there must be an effective system in place to monitor and manage the performance of the facility equipment and systems. Close monitoring of facility equipment and systems offers opportunity to intervene in the failure cycle (P-F curve) by early identification of signs of equipment degradation that can be addressed ahead of the failure (see Figure 4.3: PF Curve – Equipment Failure). If these hints are not acted upon in a timely fashion, equipment failures will result, often costing significantly higher sums of money to repair or replace.

6.7.1 Operator Surveillance Rounds

In Chapter 5, section 5.5.5, we introduced operator surveillance rounds. Operator surveillance rounds are a key element of facility operations and it is now an appropriate time to expand on the details. Special focus should be given to the optimization of operator surveillance rounds so that the time expended by operators during the round is spent as effectively and efficiently as possible. Operators should focus on only collecting equipment variance data and tailor their rounds to emphasize critical equipment over lower priority equipment.

6.7.1.1 A Good Guide to Operator Surveillance Rounds

A number of good practices and activities that should be factored into an operator surveillance round are as follows:
- A good round starts with good communications to operators, maintenance and FI&R teams – ensure that there is an effective communications protocol between these key facility teams;
- Operating procedures must be followed;
- Each round should follow the following principles:
  - clearly documented;
  - proactively reported through surveillance round check sheets;
  - round check sheets should be dated and signed off;

- All changes must be documented through the management of change process;
- All operators should be knowledgeable about their areas, including equipment and processes;
- Experienced operators should use their senses when it comes to executing their rounds, including smell, sight and touch or feel;
- Look for equipment variances, such as equipment degradation, for example drips, signs of smoke and equipment overheating;
- Focus on the known trouble spots or bad actors first;
- Check all equipment on the round.

6.7.1.2 Housekeeping During Operator Surveillance Rounds

It is important to ensure the facility is clean and tidy. A clean site enables far more effective monitoring and inspection efforts to be performed. As such, equipment variances are detected and corrected far more efficiently. Some key points to consider in housekeeping efforts for the facility during operator surveillance rounds are noted here:

- Wash up oil spills;
- Keep pump plinths clean and tidy;
- Ensure everyone is accountable for their own waste (operations, maintenance, FI&R, and contractors);
- Check that barricades are in place when required at all times;
- Keep control rooms clean and tidy;
- Check that chemicals are stored correctly;
- Ensure that waste drums are correctly labelled;
- Segregate different types of waste as required.

6.7.1.3 Visualization of Facility Inspection Rounds

One of the FIEM's key principles is to facilitate an environment where a risk-centered culture (RCC) can grow. One of the key enablers to help make this happen is to ensure that as much as possible facility equipment and systems have clear visualization of their process parameters.

How do you know that a pump is performing within its OE if there are no indications on the process parameter gauges (pressure, temperature flow, etc.)?

In addition to the equipment maintenance and operating card (EMOC), which lists the key process parameters, and the operating envelope, information located beside each piece of equipment, each of the process parameter gauges should be color coded or highlighted to ensure it is clear that the equipment is healthy. These gauges may be color coded to show the OE ranges and clearly identify the safe operating zone. It is imperative to identify the unsafe operating zones. By doing this, steps towards the creation of a risk-centered culture are created, to hold all facility groups including operations, maintenance, FI&R, facility management, engineering, etc., accountable to ensure the health of the facility equipment. If variances are identified, they should be clearly evident based on the visualization technique and can therefore confidently be reported by the individual, whichever facility group they belong to.

Figure 6.4 shows a photo of facility equipment gauges that have been color coded to show this concept on a facility. It is therefore easy to identify the OE and the performance of facility equipment within the OE.

6.7.1.4 Quality Assurance: Compliance of Operator Surveillance Rounds

Given the importance of the operator surveillance rounds, it is necessary to introduce a quality control measure, in this case an assurance health check process, which ensures surveillance rounds are executed in a timely manner

Figure 6.4 *Visualization of facility equipment performance.*

Quality Assurance: Compliance of Operator Surveillance Rounds				
Date:	**Facility Process Area:**			
	Equipment Audited:			

	Yes	No	NA	Comment
Operator training is complete for surveillance rounds and variance reporting				
All Facility equipment have a Equipment Maintenance and Operating Card displayed beside the equipment				
Equipment Maintenance and Operating Card details have been validated (Criticality and OE)				
Equipment variances are being captured in database				
Variances have all been assigned an owner and are being closed out				
Variances are being closed out in a timely fashion				
Surveillance rounds are being undertaken and recorded as per the schedule				
Critical equipment are clearly identified and marked on the facility, (e.g. pump plinths are painted red for high critical equipment)				
The surveillance round route is clean and tidy				

Signed Off: _____   Signed Off _____
                    Auditor   **Operations Supervisor**

Figure 6.5 *Audit compliance of operator rounds.*

and comprehensively. Figure 6.5 shows a sample audit template that can be used to achieve this purpose, which is aimed at assessing the compliance of operator rounds. It is a simple but effective tool that gets to the bottom of how well surveillance rounds are performed quickly. This tool may also be applied to maintenance and condition monitoring surveillance rounds. Quality assurance is a key feature of the FIEM and shall be presented in Chapter 7.

6.7.2 Equipment and System Variances

Equipment and system variances may be defined as "a deviation from the equipment 'normal operation.'" Variances may be in the form of a drop in overall equipment performance, such as a reduction in the discharge pressure of a pump, or a physical variance such as a leaking mechanical seal or a noisy cavitating pump.

Variances are a central concept in the FIEM and facility operations in general. Variances should be raised when equipment performance parameters drop in general but in particular if they drop below the operating envelope (OE). Variances should be raised as soon as they are identified on the facility but specifically they are recorded as an output of operator surveillance rounds that we discussed in Chapter 5. Each piece of equipment and system is assessed during surveillance round and any signs of performance degradation or *variance* to the equipment normal operating performance are noted. During the surveillance round the operator may consult the Equipment Maintenance and Operating Card (see Figure 5.6) to assess if there is a deviation from the OE. If this is found to be the case the operator records the variance.

Equipment variances are usually tracked in a database that is stationed in the Operations Control Center. The variance database can be a simple user-friendly software package that is running 100% of the time so that operators and other facility groups can log in and update facility variances they have noted during their rounds. These variances are logged against the facility equipment register, which includes the critical equipment list. The database can then issue a simple report to list the variances identified for various facility groups to address. Normally these variances would be discussed in a daily meeting with each of the functional groups and ownership assigned to address them.

6.7.3 Equipment Status Board

In order to keep track of the status of the potentially vast amount of facility equipment and its performance, it is often the case that a facility equipment status board is used. This board can take many forms and is a common method used to monitor facility equipment.

The equipment status board usually takes a central position in the control room, in view of the operations morning meeting team, which will be discussed in section 6.8. The board contains an exhaustive list of all facility equipment on that particular area of the facility, usually by equipment number. The equipment list is arranged hierarchically by criticality ranking, and a status indicator is assigned to each item. The indicator may be color coded, such as:

- Green = Equipment healthy;
- Amber = Equipment running with concern;
- Red = Equipment is out of service.

The equipment status board is reviewed daily and the indicator is adjusted on an ongoing basis each day.

An example of a facility equipment status board is shown in Figure 6.6.

| Facility Equipment Status Board | | | | | | | | | |
| HEAT EXCHANGERS | | | | PUMPS | | | | | |
Area 1	Status	Area 2	Status	Area 1	Status	Area 2	Status	Area 3	Status
X123		X142		P112		P131		P150	
X124		X143		P113		P132		P151	
X125		X144		P114		P133		P152	
X126		X145		P115		P134		P153	
X127		X146		P116		P135		P154	
X128		X147		P117		P136		P155	
X129		X148		P118		P137		P156	
X130		X149		P119		P138		P157	
X131		X150		P120		P139		P158	
X132		X151		P121		P140		P159	
X133		X152		P122		P141		P160	
X134		X153		P123		P142		P161	
X135		X154		P124		P143		P162	
X136		X155		P125		P144		P163	
X137		X156		P126		P145		P164	
X138		X157		P127		P146		P165	
X139		X158		P128		P147		P166	
X140		X159		P129		P148		P167	
X141		X160		P130		P149		P168	
Green		Running							
Yellow		Minor Problem - Monitoring							
Red		Out of Service							

Figure 6.6 *Equipment status board.*

6.7.4 The "Operations Meeting"

In order to ensure alignment between the operations, maintenance and FI&R team, and to facilitate knowledge sharing, highlight variances, etc. a meeting should be held. Usually this meeting would be held in a central location on the facility such as the Operations Control Center. It is often referred to as the "operations meeting" and is held first thing in the morning so that all parties can prepare themselves for the priorities of the day, factoring in the activities from the previous night shift.

The operations meeting covers a number of agenda points but with respect to integrity management it focuses on how the facility is performing and what actions need to take place to address any variances and other related concerns such as safety and resources, etc. The meeting should carefully review any new facility equipment variances that have been generated during the facility night shift and assign ownership to resolve them and follow up on variances already logged into the variance database.

The facility equipment status board may be displayed in the Operations Control Center and provides a central input into the operations meeting. During the meeting the board would be reviewed and adjusted accordingly with updates from the various facility groups, such as maintenance, FI&R and operations. Equipment variances would also be updated during the course of the meeting.

A typical agenda for the operations meeting may include the following:
- Review the operations log;
- Review of the production targets;
- Overview of Facility equipment performance:
  - Focus on performance of critical equipment;
  - Review of new and existing equipment variances (output report from the Variance database);
- Review and update of the equipment status board;
- Maintenance work status (new tasks, ongoing, delayed, due to handover);
- Review of opportunities for improvement.

6.7.5 Management of Facility Variance

Figure 6.7 illustrates a typical variance management process. The first step in the process is about identification of variances on the facility. Essentially operators on their surveillance round would review each item of facility equipment on their round and compare the operating conditions to the Equipment Maintenance and Operating Card (EMOC). Operators are focused on identification of variances. On identification of a variance at the facility, operators would record it during their round in the field. Variances from the surveillance round would then be uploaded to the variance database in the operations control center.

The next phase is about managing variances. The operations supervisor would then issue a report from the last 24 hours of variance activity

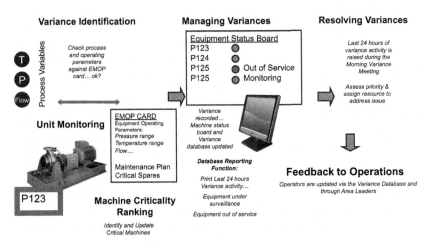

Figure 6.7 *Management of facility variance.*

as an output of the variance database in preparation for the operations meeting. During the operations meeting each variance is reviewed and assigned a priority and owner to address it and drive it to completion. The owner is usually a maintenance technician from the maintenance team.

Thirdly, attention is turned to resolving variances. It is often the case that variances may be updated into the CMMS as work requests, depending on the effort required to address them.

It may be the case that initially there are a high number of variances identified, especially if this initiative is new to an organization. In this case, a good value adding strategy would be to appoint a small dedicated team to focus entirely on addressing equipment variances. This strategy may be applied as a temporary arrangement until the number of variances generated reduces to a manageable level.

Finally, once the work is done and the variances are resolved, it is checked by the originator. The variance record in the variance database is then signed off as closed. A record of the variance is updated along with the details and feedback to the originator. It is important to ensure that the variance is properly closed out in the database for traceability of equipment history. Variance records form an essential part of equipment history and therefore must be updated in the CMMS in order to ensure equipment histories are complete.

The illustration in Figure 6.8 is of a typical variance database entry screen. The database is designed with a simple entry screen that shows a selection bar for equipment on the right hand side. Once equipment is selected, basic information is presented, followed by a brief history of associated variances and corresponding comments.

The database may also be equipped with a set of user-friendly basic reports that include previews of variances over different durations and equipment histories. These reports may be used during the operations meeting as an agenda item to facilitate the review of the equipment variance status.

6.8 MANAGEMENT OF FACILITY VARIANCES WORKFLOW

A simplified workflow process that details out the management of facility variances is shown in Figure 6.9, which is an extension of Figure 6.7. The workflow starts off with the identification of equipment variances via

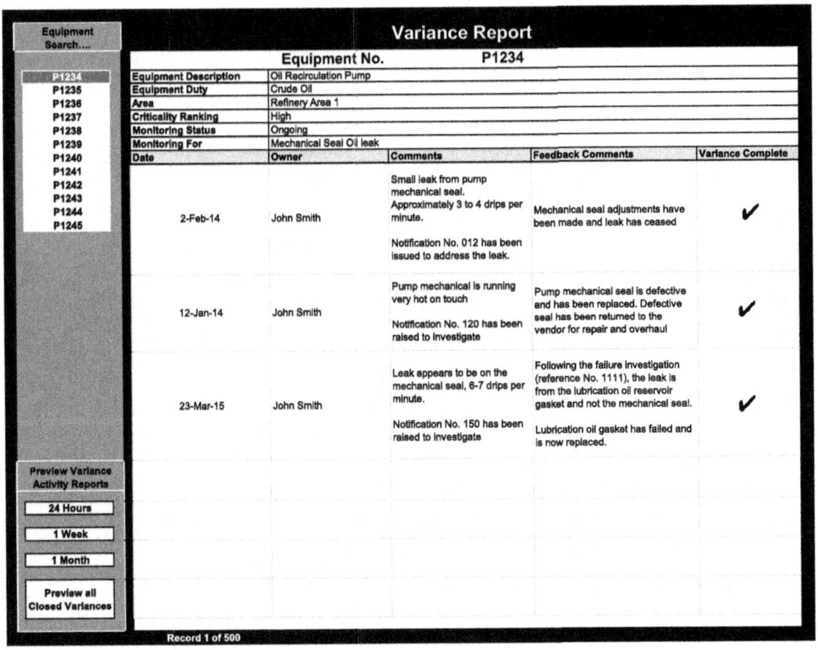

Figure 6.8 *Facility variance database.*

operator surveillance rounds. It is noted that the identification of variances may also be achieved through the maintenance group. This is done through condition monitoring surveillance rounds or even through other facility groups including facility management and site contractors. The initiative aims to drive risk-centered culture throughout the entire facility. The variance process is a key enabler for this to happen by creating an awareness of equipment reliability performance and variances and an opportunity for a collective effort to resolve them.

Once variances are identified they are captured in the variance database. This activity can be directed through certain groups at the facility such as maintenance or operations or performed by all groups on the facility depending on the preference of the facility management teams. Traceability can be achieved by developing a user identifier log-on for the database. In order to ensure that there is an acceptable standard of quality as the variance is updated in the database, it is important to ensure the facility organization is sufficiently trained in the variance process and database system.

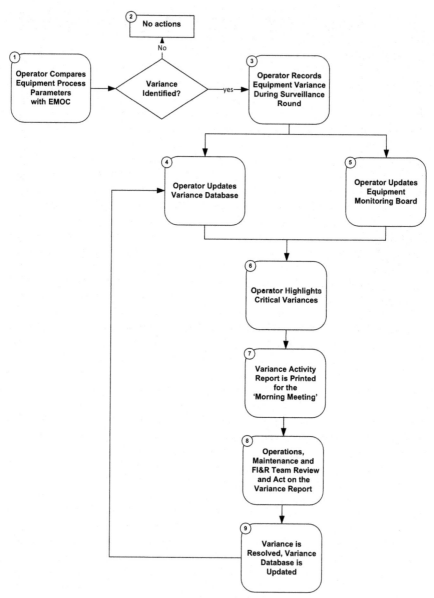

Figure 6.9 *Management of facility variances workflow.*

The workflow moves on to prioritization of the variances in terms of criticality. Since each variance is linked to facility equipment, one of the straightforward ways to assign priority is to adopt the criticality rating of the equipment in question regardless of the severity of the variance. However, this may depend on the specific situation on the facility.

The next focus is to act on the variance. During the morning meeting each variance is discussed and resources are assigned to address each one, starting with the most critical ones. As the variance is resolved it is important for the originator to be satisfied that the work has been done to a satisfactory standard and therefore check it at site. The variance can then be updated in the variance database and signed off as complete.

CHAPTER 7

Integrity Supporting Processes

Contents

7.1 INTRODUCTION

We have covered the facility segment of the Facility Integrity Excellence Model (FIEM), which includes maintenance, operations and FI&R. Chapter 7 presents the second segment of the FIEM, *integrity supporting processes*. Integrity supporting processes are paramount components of the FIEM and facility integrity management programs. Integrity supporting processes cement the core segments of FIEM together, which include facility personnel and the technical elements associated with the facility. They include five critical processes: namely, management of change, quality assurance, incident reporting, management of knowledge, and management review.

In order to operate effectively, the FIEM integrity supporting processes must be able to integrate with each other. Information must be able to

Facility Integrity Management
http://dx.doi.org/10.1016/B978-0-12-801764-7.00007-3

flow freely between the various FIEM workflow processes as it is intended to, without constraint or hindrance. It is important to ensure that the flow of information between the numerous FIEM workflow processes is also clear and robust. FIEM processes require up-to-date information in order to function properly.

The information flow may be in various forms, including human interaction and electronic information. Having clear and robust information flow issued by one facility group and received by another also encourages collaborative working between the facility groups and, in the process, breaks down the barriers that create silos.

7.2 MANAGEMENT OF CHANGE

By carrying out modifications to existing facility equipment and systems in an uncontrolled manner we become exposed to potential integrity incidents. We may refer back to the case study in Chapter 2, section 2.6.1 (Case Study – ConocoPhillips, Humber Refinery). One of the main causes of the explosion on the Humber Refinery was poor management-of-change (MOC) processes and procedures.

The key objectives of any MOC process are to ensure that new risks are not accidentally introduced to the facility and that the current risk profile is not adversely affected. Management of change or MOC identifies the requirement to evaluate and approve changes before they are made at the facility. MOC also provides a tool to document all changes at the facility, and review and approve them to ensure that all changes are installed as designed without compromise before they are implemented for a safer and reliable operation (Figure 7.1).

The MOC process provides a control barrier (see Figure 2.6: Swiss cheese model – Barrier Analysis) in order to further safeguard the integrity of the facility against a variety of potential mishaps, which may include poor or improper design; incorrect construction standards; inappropriate materials of construction; poor execution, etc.

The MOC process should also be extended to the facility organization. This is because a change in the facility integrity organization can adversely affect the integrity of the facility. For example, if new equipment or specialist equipment are introduced to the facility, there is a different resource requirement. If we ensure that the integrity organization follows the MOC process, we will have new confidence that our facility is right sized in terms of competence and number to support the facility integrity effort.

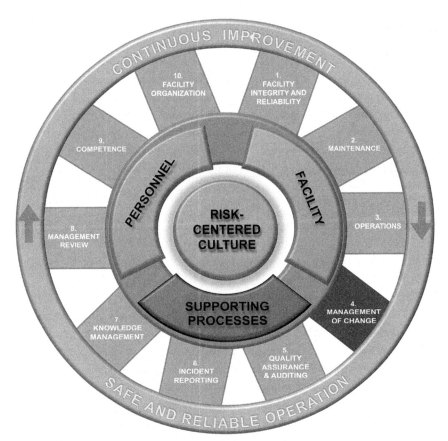

Figure 7.1 *Supporting processes: Management of change.*

A MOC process typically deals with the following changes to the facility:
- Changes to the facility process design;
- Introduction of new equipment and systems;
- Integration of existing facility process equipment with new facility process equipment;
- Changes to or new maintenance tasks that impact the current equipment and systems design and operation of the facility;
- Decommissioning or mothballing existing equipment;
- The facility organization.

The potential impact of changes on facility integrity must be detailed, monitored, closed out and sufficiently communicated to the relevant groups at the facility. The MOC process must be administered in an effective and timely manner.

Management of change refers to all changes at a facility, which can be difficult to administer with a single MOC procedure because changes may differ considerably in size and complexity. It is important that all changes follow the same process in review and approval; however, the level of scrutiny and management approval may differ based on the potential impact and urgency of the change. Since all changes do not need the same extent of evaluation and approval, the MOC process and procedure must be flexible enough to address each and every change. This may be done by categorizing each change and evaluating it to the appropriate level of scrutiny and approval.

7.2.1 Types of Change

It is important that all changes that may have an impact on the integrity of the facility are considered by the MOC process, not just the ones you can see on the facility such as the physical ones. As such we can broadly categorize changes as follows:

- **Physical (and Software)**
  - Modifications to equipment including materials or equipment specifications;
  - Deviations from design intent (nonconformances, basis of design);
  - Changes to current equipment suppliers;
  - Modifications to process safety system such as protective systems;
  - New facilities and associated tie-in to existing facilities;
  - Changes to maintenance activities;
  - Advances in technology.
- **Facility Process Related**
  - Changes to facility process feed stock;
  - Changes to the facility process design;
  - Changes to the operating envelope.
- **Organizational**
  - Changes to current job roles and responsibilities;
  - Changes to the organizational structure (new positions added or existing positions deleted);
- **Documentation**
  - Changes to plans, procedures, and drawings, etc.;
  - Changes to equipment or material standards and specifications;
  - Changes to operating envelopes;
  - New statutory and other regulations.

These lists are nonexhaustive and only intended to provide a guideline of some of the different types of change that the MOC process will have to deal with. The key point is that management of change deals with many aspects over and above physical changes to the facility.

7.2.2 Management of Change Workflow

The MOC workflow process is illustrated in Figure 7.2. The process essentially comprises the main steps described in the following subsections.

7.2.2.1 Identify the Need for Change

The need to identify change may come about from a number of different sources. We can refer back to the different types of change in section 7.2.1 to help visualize some of these circumstances. These may be due to new projects that must integrate with the existing facility, changes to the organizational structure, new technology that has been introduced to improve the process, etc.

7.2.2.2 Prepare MOC Proposal

In order to provide a detailed understanding of the change so that it can be properly assessed, it is important for the change originator to develop a detailed proposal. The proposal must provide a comprehensive account of the change, including definition of the current state, an account of the background to the change and details of the actual change. The MOC proposal takes the form of a structured document and should cover the following key topics:

- Detailed description of the change, including relevant documentation to fully describe any technical elements and the rationale behind the change, such as piping and instrumentation diagrams (PIDs), layout drawings, calculations, technical design basis, etc.;
- Any health, safety or environmental (HSE) impact, including any HAZOP or PHA studies that have been performed;
- Updated procedures, including operating, maintenance and FI&R;
- Training requirements;
- Changes to facility resources and the facility organization;
- Communication plan for the change – how will the facility organization be informed of the change?

7.2.2.3 Raise the MOC Request

The MOC may be raised through an MOC software program or alternatively by a process that provides a structured review and approval process of the MOC proposal. The latter may involve a meeting with a cross-section

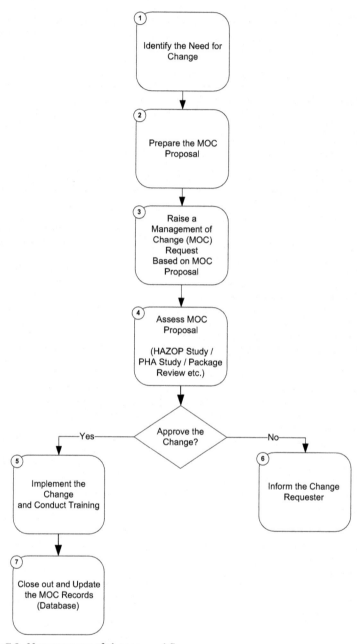

Figure 7.2 *Management of change workflow.*

of attendees to present, discuss and approve the change. One of the key requirements of the MOC system is to ensure comprehensive tracking of the change so that no changes are forgotten and the basis of approval is adhered to.

The MOC tracking system also ensures that MOCs, including temporary MOCs such as process trials, for example, are closed out in full, with all actions signed off. The main purposes of the MOC tracking system are to ensure the MOC does not get lost in time and that it is closed out in a timely fashion.

7.2.2.4 Assess the MOC Proposal

The assessment of the MOC proposal should be carried out by suitably qualified facility personnel and includes appropriate representation from the relevant technical disciplines as required. The MOC proposal assessment may take the form of independent reviews or via a cross-sectional forum of appropriate facility representatives. The assessment should evaluate the MOC proposal document in terms of completeness, accuracy and substantiation of any assumptions made. The assessment must ensure that all possible impacts of the change are assessed and that any interface with other facility processes and systems are understood and assessed as well.

Process Hazard Analysis

In some cases it may be appropriate to use specific tools to assess process hazards on the facility. One such tool that is used extensively throughout the oil and gas and petrochemical industries is a process hazard analysis (PHA). PHAs are used to identify, evaluate and control significant hazards in the facility. These hazards have potential to cause explosions, fires and the release of harmful substances.

PHAs employ a structured and systematic approach in the analysis of process hazards, by deploying a multidiscipline team to critically review hazards and document the output. The PHA report is used as a basis for training of facility operations and maintenance personnel and also for facility emergency planning.

PHAs are usually employed without fail when there is a new facility, during a cyclic review of an existing facility, and during the decommissioning of existing facilities. Cyclical reviews are performed usually over 2- or 5-year intervals and aim to assess the entire facility including any modification made since the last cyclic review. The new facility PHA is used as the baseline PHA and serves as a foundation for future cyclic PHA reviews.

The PHA process involves the following key steps:

1. Identification of process hazards;
2. Analysis of the consequences of the process hazards;
3. Evaluation of the process hazards;
4. Development of recommendations to control the process hazards.

MOC Prioritization

One of the underlying principles of the FIEM is to ensure that precious facility resources are assigned to meet the work requirements appropriately. There is no exception when it comes to MOC. Assigning priority also extends to the review and approval resources required. The highest priority MOCs may require approval from the facility management team, whereas the lower MOCs may be approved by facility area management, for example. MOC proposals are assigned a priority during the assessment stage. There are many ways to assign priority; however, it is often pertinent to assign priority based on the associated risk.

7.2.2.5 Implement the MOC

The implementation of each and every MOC must be executed in line with the agreed proposal after it has been assessed and potentially amended. This is absolutely essential because any deviation of MOC during its implementation may result in a different impact associated with the modification, which may give rise to unforeseen risks. Any variations during the implementation of the modification should be highlighted and be subjected to further review. The MOC proposal document should also be updated to reflect these changes.

7.2.2.6 MOC Safe Start-Up

Once the modification has been successfully implemented in line with the approved proposal, the facility or part of the facility can be restarted. However, given that there has been a change made to the facility, it is important that a process be used to ensure the facility start-up is done safely.

One way to achieve this is to carry out a *pre-start-up safety review* (PSSR) process, which we discussed in section 6.2.1. This is to ensure that the facility organization is satisfied that the modification has been completed as designed in compliance with the relevant codes and standards so that the integrity of the facility is not compromised whatsoever. The PSSR should also ensure that training has been completed and that relevant procedures and operating instructions have been updated.

7.2.2.7 Close Out and Update the MOC Records

Once the change has been implemented and the facility has been started up after the PSSR, the MOC can be closed out. During this process the MOC documentation must be updated and signed off, all relevant facility personnel must be informed of the change, and training carried out. The MOC tracker database can then be closed out. It is important that the facility MOC records are accurate and comprehensive, in the event that future facility changes and improvements are made that require a review of historic information covering facility changes.

7.2.3 Facilities MOC Procedure

MOC accounts for any change whether it is temporary or permanent to the facility organization, which may include equipment, software, materials, personnel and procedures that may have a detrimental effect on the integrity of the facility or the company's business objectives. MOC effectively subjects all changes to a risk assessment procedure, which is performed by appropriately qualified and experienced personnel.

The MOC procedure must be comprehensive and clear in order to administer the MOC process. It intends to provide guidance to control and document all changes that may affect the integrity of the facility. All facility groups must be conversant with and follow this procedure.

The MOC procedure clearly defines the approval process and approvers and puts measures in place to ensure the implementation of the change is executed correctly as agreed. The procedure also mandates that all relevant MOC documentation such as drawings, procedures, studies, etc. are well documented and properly stored.

The procedure also ensures that the implementation phase of the change includes consideration for safe start-up of the facility. This is to confirm and provide confidence that the change was made as designed in the MOC proposal and as approved.

The main objectives of the MOC procedure are as follows:

1. To protect the facility equipment and systems from unauthorized changes that may jeopardize the integrity of the facility;
2. To document, track and provide an auditable record for each and every change at the facility;
3. To ensure that the process of developing, reviewing, approving and implementing changes at the facility is robust as well as smooth and as cost effective as possible.

The MOC procedure is intended to be in place for the management of all changes throughout the facility life cycle. It is important that the MOC procedure be communicated and understood by all the facility organization. Changes made to the procedure should also be subjected to the same rigorous MOC process.

7.2.4 Management of Change Training

Management of change affects the whole facility organization. This is because changes may be introduced to all of the facility groups including operators, maintenance technicians, inspectors, engineers, etc. It is therefore important to ensure that a training program is developed to meet these needs.

The MOC training program should ensure the facility organization understands what a change is and what the potentially serious consequences are if the MOC process is not followed. Once the initial MOC training has taken place, each employee should also undergo refresher training in order to reiterate the key principles of MOC at suitable intervals.

7.3 QUALITY ASSURANCE

7.3.1 Introduction

In the context of the FIEM, quality plays a key role for each and every element associated with the management of facility integrity. Quality assurance can be defined in a number of ways. It refers to the end product in terms of meeting the company's and customer's sale specifications; it also refers to quality of workmanship for maintenance overhauls, quality of workflow processes, quality of inspections performed, quality of training, and so forth (Figure 7.3).

7.3.2 Quality Management System

A quality management system (QMS) focuses on achieving the quality policy and quality objectives that drive to meet the company and customer requirements. The QMS is articulated through the facility integrity organization: its policies, procedures and processes that are required in order to successfully achieve quality management of the facility.

All work that is performed at the facility including capital projects, maintenance, engineering design, and procedures, as well as the feedstock and end products, are all subjected to the QMS requirements. If we do not carefully manage and implement the company quality management system

Figure 7.3 *Supporting processes: Quality assurance and auditing.*

we are exposed to potential failure of our facility integrity management system.

Poor quality control can lead to higher overall operating costs; on the other hand, there is no point in having overachieved the highest quality specification of the end product if we do not see the return on products sold. Quality assurance must be balanced with cost. Quality assurance can be expensive and must be budgeted accordingly. We need to achieve the right level of quality for the facility at the lowest cost.

A quality management system also provides transparency in the day-to-day operations of the facility. This is important for a number of reasons. QMS enables the evaluation of the current performance for each and every facility group, including third-party contractors and the supply chain, with the aim of ensuring a consistently high quality of work and ultimately improving performance as well as providing a strong foundation for compliance with legal and regulatory requirements such as ISO (International Organization for Standardization) [7.1]. Quality management is an essential element of the FIEM and must be accomplished effectively.

Quality management encompasses both quality control and quality improvement. Ultimately, quality management plays an integral role in the FIEM principle of continuous improvement which addresses each FIEM element. A prerequisite to the effective management of quality is to have a robust audit and review workflow process.

7.3.3 Quality Management of Maintenance Work

It is worth spending a moment to consider the particular application of quality management to maintenance work. This is because there is a high potential for premature equipment failure and lowering of the mean time between failures (MTBF) as a result of maintenance work that does not meet the appropriate quality assurance standards. When equipment is correctly maintained using the right consumables, method statement and spare parts and materials, it performs its intended duty more effectively and therefore operates with a longer MTBF. The quality assurance of maintenance work involves checking that after each work order is complete, the equipment meets its design specification and ultimately the maintenance work is performed properly.

This involves ensuring that the maintenance team follows the written procedures on how to perform the maintenance work. The procedures should include steps to check that certain critical activities are carried out correctly: for example, check and confirm the correct gasket has been installed, check that the correct bolt torque is applied, or check the correct materials of construction are being used.

7.3.4 Audit and Review

Quality audits are necessary to verify the conformance of FIEM processes and to assess how well the processes have been implemented. It is important that a robust audit program be developed and implemented for facility integrity that addresses each and every element of the FIEM.

The competency of the quality personnel administering the program should also be carefully evaluated. We can have the best audit program but if the personnel running the program are incompetent and inexperienced, then we will not see the benefit. We shall discuss competence and the facility organization in detail in Chapter 8. Guidelines for audits and audit programs are highlighted in ISO standard 19011:2011 [7.2]. The key objectives of the FIEM audit and review process are as follows:

- Develop and implement an Audit Program to effectively deliver the requirements of the FIEM audit and review process;
- Ensure compliance of the facility quality management system objectives with the FIEM processes and systems;
- Highlight any quality concerns and provide recommendations for improvement;
- Provide a foundation to drive for continuous improvement.

7.3.4.1 Development of an Audit Program

The objective of the audit program is to provide assurance that all of the processes that make up the FIEM are functioning effectively and as intended. It provides a structured approach to audit facility integrity processes and to identify any compliance concerns and opportunities for improvement.

The audit is kicked off with an introductory meeting by the auditee and audit team. During this meeting the scope of the audit and its objectives are presented and discussed. Audits are executed against audit checklists, which are developed in advance. They may be conducted by an individual or team of auditors based on the scope of the audit. The audit systematically works through the audit checklist. As the audit progresses, corrective actions are noted where nonconformance or opportunities for improvement are identified. Corrective actions are assigned owners and time scales for completion.

The audit is wrapped up with an exit meeting by the auditee and audit team to discuss the findings. A short period after the audit exit meeting, a detailed audit report is issued that discusses the corrective actions and opportunities for improvement.

Audits may be categorized as external audits and internal audits. External audits are carried out by an independent body that resides outside of the organization being auditing. External audits tend to report to the company's shareholders and provide an account of company compliance, usually with respect to national or international standards and regulations. External audits tend to be more formal and examine each aspect of the system being audited without restrictions. External audits are usually performed on an annual basis.

The facility management team tends to set the objectives and scope of audit for internal audits, highlighting certain focus areas. Internal audits reside within the facility organization and report to the facility management.

They provide a useful tool to review and improve the facility's workflow processes, including risk management procedures. Internal audits can also review the operating effectiveness of the facility's workflow processes to make sure they are working properly. There can be many different types of internal audits, which are scheduled on a relatively frequent basis, such as quarterly.

7.3.4.2 Quality Nonconformance

During the audit, nonconformance to company standards, legislation and facility procedures may be picked up; these issues must be addressed promptly. In order to ensure that there is an effective approach for addressing nonconformance, it is important that all issues related to nonconformance be carefully reviewed and approved by competent and experienced personnel staff. In the case of major nonconformance, the proponents of particular standards should be consulted.

7.4 INCIDENT REPORTING

An incident may be defined as an unplanned event that has resulted or may result in harm to facility personnel or damage to the facility. Incidents and potential incidents (near misses) will happen again unless the underlying root causes are identified and corrected. By thorough investigation and reporting of all incidents and potential incidents, we can continuously improve the performance of the facility. This can lead to step changes in integrity performance as well as safety performance.

An incident as we have defined it is an unplanned and undesired event that has taken place and has hindered the completion of a certain task. Incident reporting also includes potential incidents or *near misses*. Incidents are preventable and by reporting them and asking ourselves why they have occurred, we can prevent them from reoccurring (Figure 7.4).

7.4.1 Near Misses

A near miss may include any situation at the facility where there is potential for an accident to occur but there is no actual harm to facility personnel or the environment or damage to the facility. Near misses may include the following situations:

- A failure of a barrier of protection (before the final barrier that results in the incident) (see Figure 2.6: Swiss cheese model – Barrier analysis);
- An unsafe condition, such as a slippery walkway or loose hand rail;

Figure 7.4 *Supporting processes: Incident reporting.*

- An unsafe act, such as a maintenance activity being performed with incorrect PPE;
- Unsafe tools or equipment, such as defective hand tools for maintenance or equipment without proper certification or certification that is out of date;
- Exceeding an operating envelope (see previous Figure 6.3: Facility operating envelope);
- A loss of containment of a process fluid.

Moving the facility organization to proactively identifying and reporting near misses is a fundamental step forward towards a risk-centered culture. It is a good way to get the organization to discuss integrity and safety in general on a daily basis.

7.4.2 Incidents

We have defined an incident as an unplanned event that has resulted or may have resulted in harm to facility personnel or damage to the facility. An incident must be reported swiftly in order to minimize the response time, so that medical care or an emergency response can be administered in a timely fashion. It is common place for companies to operate an *Online Incident Reporting System* that enables proper prioritization and tracking of incidents (and near misses). The incident reporting system captures the essential first-hand information from the source – the incident reporter. This information is then fed through the appropriate channels in order to ensure the right level of response is provided. For serious incidents, reporting should additionally be made to the facility management team and/or safety manager, depending on the nature of the incident. All incidents should be reported within 24 hours of the incident occurring.

It is important for an incident reporting procedure to be developed and implemented for the facility organization in line with the FIEM. The procedure should define the requirements for incident and near-miss reporting, the investigation process and action development and tracking. The procedure should also enable each incident and near miss to be prioritized so that the appropriate level of resource can be assigned.

The incident reporting procedure should be integrated with the root cause analysis workflow; refer back to the matrix in Figure 4.14 (Failure investigation matrix). This is because incidents and near misses should be analyzed to understand their root cause in order to prevent reoccurrence.

7.4.3 Reporting of Incidents and Near Misses

A record of all incidents and near misses that have taken place at the facility should be captured within the online incident reporting system. All incidents require an incident investigation report to be prepared, even more so if there is an incident related to a high potential consequence. It is essential that all facility personnel be aware and conversant with the use of the online incident reporting system or a similar tool for the reporting of incidents.

All incidents must be investigated thoroughly, based on facts and accurate information that have been collected during the incident investigation. In order to be consistent in the approach to incident investigation, the failure investigation matrix in Figure 4.14 can be updated to incorporate the different incident priorities.

Utilizing the online incident reporting system, it is often a worthwhile exercise to analyze historic incidents in order to identify and develop trends in a facility. The output of this exercise may enable the facility management team to focus on improvements to the various facility processes. For example, the trends in facility safety performance may point towards enforcement of proper PPE, or failure of equipment in certain areas of the facility may point towards problems associated with the particular facility systems rather than facility equipment.

7.5 MANAGEMENT OF KNOWLEDGE

An effective knowledge management system is a fundamental element of the FIEM. Reliable and complete information provides a solid foundation for continuous improvement and credible decision making. The main objectives of the management of knowledge element of the FIEM are to ensure that facility integrity data are recorded, reviewed and kept up to date for the relevant facility groups as required (Figure 7.5). Facility integrity information can then be confidently communicated between the various facility groups.

The FIEM management of knowledge process is made up of a number of difference components. The first is through the implementation of a reliable and effective means of collecting and recording integrity information. Integrity information is required in order for all of the facility integrity groups to operate effectively.

The management of knowledge process should have an effective means to transfer integrity information to and seek feedback from the numerous facility groups. A communications plan should be prepared to ensure that the information flow between each facility group is clear and that information flows between the groups.

The communications plan should also detail the communication means, whether it is by email, via a meeting, on a display board, etc., along with the information required and the frequency of the communication.

Knowledge management must also enable effective and reliable storage and maintenance of information for the various facility groups' databases or document management systems. In the interest of consistency and to ensure that facility integrity data is "clean," with no duplicated records, there may be a central "master" database at the facility. This is usually the computerized maintenance management system (CMMS). Facility group specific

Figure 7.5 *Supporting processes: Knowledge management.*

databases may also be integrated to better serve the specific needs of the facility integrity management system, or the FIEM.

It is also important to ensure that regular audits are conducted against preset key performance indicators (KPIs) in order to monitor the effectiveness of knowledge management at the facility to ensure a healthy flow of information throughout the facility.

7.5.1 Document Management

FIEM relies on the timely and effective transmission and maintenance of facility integrity documentation. The effective management of facility documentation is an essential element of the FIEM. Document management systems can range from the structured filing of document hard copies in a filing cabinet to a fully blown enterprise management system.

A document management system (DMS) manages the life cycle of facility documentation. It controls how documents are created, reviewed, approved and issued. An effective DMS specifies the types of documents including the templates to be used, the storage requirements, and the access control requirements for each type of document. It also specifies the retention and maintenance requirements for each document; for example, some documentation may require retention over extended periods due to legal requirements.

The facility DMS should be flexible so that it can be tailored to the user's requirements, which may differ depending on the requirements from each facility. Some documents, may require more stringent controls due to their nature, such as legal or compliance-related documentation.

An effective DMS enables facility integrity documentation and information to be easily recorded, stored, and shared between the facility groups, which helps create an environment to promote a risk-centered culture.

A DMS is essentially a smart database that is able to effectively manage large volumes of data. This includes root cause analysis reports, maintenance equipment performance reports, EMOPs, variance reports, OEs, PSSRs, inspection reports, and others. Communication is required between all of the facility groups and their corresponding systems, including CMMS, operations and production control system, ICMS, MOC, variance database, and so forth.

An illustration of the main facility integrity databases and document management systems is shown in Figure 7.6. Each facility group utilizes these databases to assist in the integrity assurance of the facility. The integrity and management of this information is critical to the success of the facility. These facility integrity databases and document management systems provide the instruments to ensure all of the facility integrity groups, including maintenance, operations and FI&R, are managed effectively.

7.5.2 Flow of Facility Integrity Information

The management and control of facility integrity information between the numerous facility integrity groups is a complex undertaking that needs to be carefully handled. Many of the facility integrity groups maintain databases or document management systems with automated information workflow in order to cope with the high volume of information flow requirements. These systems tend to utilize a "rules-based workflow" which requires a preprogrammed set of rules to manage the flow of information

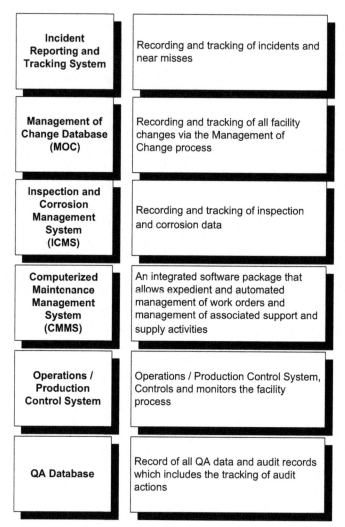

Figure 7.6 *Facility integrity databases and document management systems.*

throughout the integrity organization. For example, a maintenance work order flow would be routed through the appropriate groups, which may be operations for production planning, FI&R for inspection, work requests via CMMS, etc.

Facility data sources and databases along with the flow of information are illustrated in Figure 7.7.

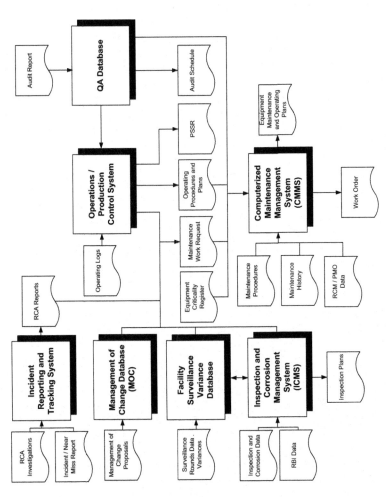

Figure 7.7 *Management of knowledge: Flow of facility integrity information.*

7.6 MANAGEMENT REVIEW

One of the critical success factors of the FIEM is the structured approach to the Plan – Do – Check – Act, or Shewhart cycle, and in particular the feedback loop (check and act). Each element that makes up the FIEM should be continually reviewed, with deficiencies addressed and improvements made. This ensures that the FIEM elements constantly operate at the highest performance level.

In this light, regular management reviews are required and executed to assess the performance of each of the FIEM elements to drive towards integrity excellence (Figure 7.8). Management reviews are usually carried out on an annual basis, given the large volume of work required to review each of the ten essential FIEM elements over the year.

The output of the management review is kept as a record and used as a benchmark to track improvements made in each of the FIEM elements. The

Figure 7.8 *Supporting processes: Management review.*

Table 7.1 Facility integrity excellence model elements reviewed

No.	FIEM Element
1	Facility Integrity and Reliability
2	Maintenance Management
3	Operations
4	Management of Change
5	Quality Assurance and Auditing
6	Incident Reporting
7	Management of Knowledge
8	Management Review
9	Competence
10	The Facility Integrity Organization

output is also used to update best practice and procedures employed within the facility organization. The management review also ensures that any performance gaps or deficiencies are identified and action owners are assigned to close them.

The Facilities Integrity Excellence Model elements that are reviewed are given in Table 7.1.

It is also worthwhile to develop the score card to include metrics that can incorporate the extent to which risk-centered culture has been implemented within the facility organization population. This may be achieved by carrying out a detailed questionnaire and/or interviews with key a cross section of the facility integrity organisation. The management review is executed by the senior facility management team and may include a third-party company executive within the review team who is not familiar with the facility. The review team should be kept to a relatively small number of people in order to be effective.

7.6.1 The Management Review Approach

The management review is an extensive review that may take several days to complete properly. There are a number of ways to go about conducting the review. The following approach provides some guidelines:

- Define the scope of the Management Review within the Facility. This may include a particular cross section of a facility process area, or a particular functional group;
- Review the appropriate facility information as provided by the facility groups;
- Conduct a site survey and make observations based on the ongoing and historic maintenance work in the area;

- Assess and note the condition of the facility equipment;
- Interview a sample cross-section of facility personnel from the population;
- Review recent integrity audit reports and check historic audit performance and status of the audit actions;
- Review the FIEM procedures and processes for completeness and extent of implementation;
- Prepare a report to detail the findings and conclusions, ensure the feedback is presented back to the facility organisation teams and acted on.

Ahead of the review, the facility group leaders should prepare a documentation pack that includes details of the subject of the management review, such as historic trending of reliability performance, near misses, availability of the facility, etc. The facility management team should also assign appropriate representation during the site survey, typically an experienced operator and maintenance technician to accompany the management review team in order to guide the team during the review.

During the review, detailed score cards for each of the Facility Integrity Excellence Model elements should be completed. The scoring

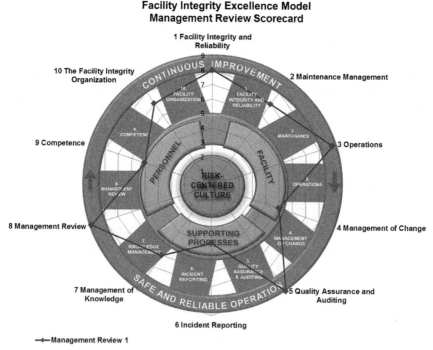

Figure 7.9 *FIEM management review score card.*

of each of the elements should be tailored to meet the requirements of each facility.

The individual FIEM element score cards are then rolled up into an overview FIEM score card. Each FIEM element is given an overall summary score and presented in a concise score card template. The score may roll up to a percentage or a total figure, but it is important that the range is the same for all of the FIEM elements in order to compare them going forward. The template may be presented in many ways. An illustration of a typical radar chart that provides a score for each of the FIEM elements is shown in Figure 7.9. This version of the score card shows an array of the FIEM element scores superimposed over the model graphic.

One of the main advantages of employing a radar chart in this process is so that future management reviews can also be superimposed on the same chart. With this method it is possible to compare progress during each management review and physically see improvements being made in certain areas. An example is shown in Figure 7.10.

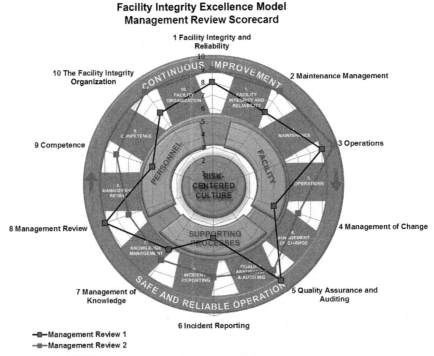

Figure 7.10 *FIEM management review score card 2.*

CHAPTER 8

The Facility Integrity Organization

Contents

8.1 INTRODUCTION

"Our people are our greatest asset."

A competent and well-trained facility organization is of paramount importance to the success of the Facility Integrity Excellence Model (FIEM). Without qualified, experienced and skillful personnel, any facility integrity program will be destined to failure.

A high-performing facility integrity organization depends on highly skilled employees who can consistently deliver quality work. These personnel must have technical skills and experience but also have key soft skills. Reliable and accurate exchange of information between the numerous facility groups enables the FIEM to function effectively and the primary way this happens is through facility personnel communicating with each other.

The facility is a fast-paced environment with many problems arising daily and running concurrently. The facility organization should therefore be able to work as a team, setting goals and targets to achieve the business objectives. Facility personnel should have problem-solving skills which are key, so that problems do not accumulate and end up in delaying the progress of certain activities. The facility organization structure should ensure that the decision making process does not present a bottle neck in the day to day operations of the facility. This can be done inadvertently by deploying team sizes that are too large reporting to one line manager, which is not conducive to a smoothly operating facility organization.

Facility Integrity Management
http://dx.doi.org/10.1016/B978-0-12-801764-7.00008-5

Creating a high-performing facility organization requires buy-in and support from the leadership team. There may be a requirement to provide funding in support of the efforts to develop the facility organization, which may also extend to providing additional support resources to drive the initiative forward.

8.2 COMPETENCE

Committed and competent facility personnel are fundamental to the success of an integrity management program. Competence refers to a combination of knowledge and skills to enable the facility personnel to act appropriately and do the right thing when faced with a diverse range of situations (Figure 8.1). Competence applies to each and every individual

Figure 8.1 *Personnel: Competence.*

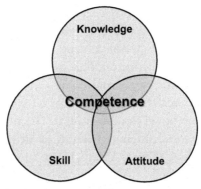

Figure 8.2 *What is professional competence?*

in the facility and at each and every stage in their careers. This is because at each level in the organization there is a corresponding level of knowledge and skill required. This also includes any third-party contractor, in addition to resident facility personnel.

Competency includes the following key attributes:

1. Knowledge
2. Skills
3. Attitude

Professional competency is a combination of practical and theoretical knowledge, cognitive skills, and the attitude of willingness to improve performance in a particular role at a certain level. An illustration of competence is shown in Figure 8.2. The illustration shows that in order to have competence, Knowledge, Attitude and Skills must overlap. A key component of this is attitude: in other words, the requirement to ensure the right values and beliefs are adopted and embraced. This goes hand in hand with having the right culture within the facility and is enabled through the principles of risk-centered culture (RCC).

It is also important to appreciate that knowledge and understanding underpin competence. Therefore, an important input to improve competence is a comprehensive facility integrity training program.

Competency levels can be defined and built into a human resources (HR) development framework. The definitions may vary but must be consistently applied across the entire facility organization.

8.2.1 The Learning Organization

During the 1940s, Edgar Dale developed an intuitive model, the "Cone of Experience" [8.1]. The Cone of Experience provides some insight into the perception of how we learn based on various kinds of audio and visual media. The model is referenced in Figure 8.3.

It suggests that how much personnel tend to remember is a function of how they encounter the information. This compares receiving verbal and visual information to participating and doing activities. The model is not founded on scientific research and should not be taken literally; however, we can take away some useful insights from it. The model suggests that, as we move from reading and listening towards participating and doing the activities we have learned, there is a step change in the impact of remembering what we have learned.

We may apply this model to the facility integrity training programs to ensure that there is sufficient consideration given to the holistic perspective of learning and not just classical classroom-based methods, which include "visual receiving and verbal receiving." The learning organization should consider a blend of verbal, visual, participating and doing learning activities. This means that, in order to be effective, FIEM training programs should include a combination of classroom-based learning with an interactive or participative element as well as a practical approach that enables the participants to perform what they have learned in the classroom as much as possible.

8.3 THE INTEGRITY ORGANIZATION

We have discussed the importance of a high-performing and competent facility organization (Figure 8.4). Competence extends beyond technical knowledge and skills; let's take a moment to consider some of the key non-technical characteristics of high-performing integrity personnel:

- They should be able to develop or propose solutions to problems encountered on the facility, i.e. be problem solvers rather than problem hoarders;
- They should be able to exercise a level of independent decision-making skills;
- They should be keen to self-develop and learn;
- They should be creative;
- They should be risk focused;
- They should embrace teamwork;
- They should be driven towards continuous improvement;
- They should possess professional and personal integrity.

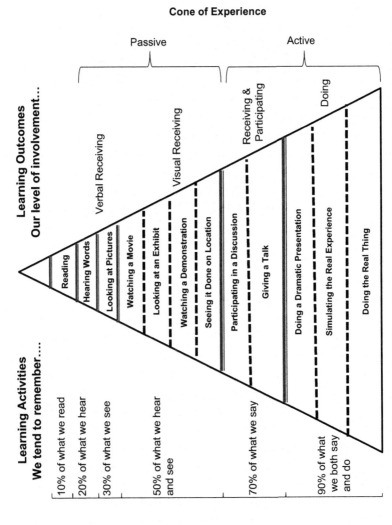

Figure 8.3 *The Cone of Experience [8.1]. (Source: Developed and revised by Bruce Hyland from materials by Edgar Dale).*

Figure 8.4 *Personnel: Facility organization.*

We have indirectly discussed a number of ways to develop facility integrity personnel with these kinds of characteristics. We have mentioned that a comprehensive training program is essential and that the training program should be holistic. This means that the training program focuses on technical skills and soft skills, as well as classroom–based and practical (doing) elements.

8.3.1 Training the High-Performance Workforce

In the continuous process of the development of high–performing facility integrity teams, the first port of call is to identify the necessary skill sets needed to support the various facility groups.

8.3.1.1 Training Needs Analysis

Training needs analysis includes the identification and evaluation of the training needs of the facility groups. Facility personnel's job role and responsibilities are used as a basis for this analysis. An exercise to identify job changes and any problems related to skills should also be completed. Each job role and associated subtasks that make up the job role are analyzed in detail and a corresponding training base of support for skills is identified.

The training plan design and development can then be executed. The design of the training plan must consider the method to administer the training, factoring in some of the concepts presented in Figure 8.3, showing the Cone of Experience.

The training plan should also consider the way the training is intended to be rolled out. It must be done in a structured way that engages its targeted audience. This also distinguishes between "first time" training and refresher training. First time training will require a more in-depth focus with longer training course durations, for example. Refresher training is about future needs and the requirement to sustain the high-performing team's knowledge and skills. There is also a requirement within the training program to bring on board or induct new employees. The training must be relevant and of high quality and involve a combination of classroom and experience-based training.

The key steps in development of the training program are illustrated in Figure 8.5.

Properly trained and performing facility integrity personnel are an integral requirement for ensuring facility equipment operates safely, effectively and efficiently. There are a number of elements associated with a holistic training plan.

8.3.1.2 Job Role and Associated Tasks

Specific job role training is important so that each and every employee at the facility understands their role and associated tasks, how to perform them well and how to troubleshoot if applicable. This training also addresses health, safety, and environment (HSE) and how to go about their particular job roles safely.

8.3.1.3 Classroom-Based Training

Classroom-based training is an essential component of a training program. It provides an environment that is conducive for learning and may include instructor-led training, interactive workshop style session, or programmed

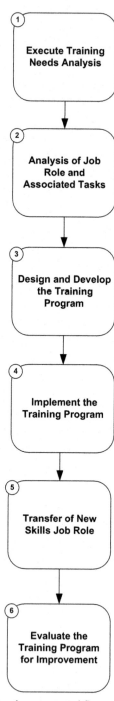

Figure 8.5 *Training program development workflow.*

instruction, among other methods. Classroom training is particularly useful for clearly explaining the numerous facility procedures and processes.

8.3.1.4 Site-Based Training

Site-based training is the practical element, or the "doing" element, that is required as part of the training plan (Figure 8.3: The Cone of Experience [8.1]). This may involve shadowing or assisting an experienced operator or maintenance technician in executing particular tasks at the facility. Site-based training complements classroom-based training, as it puts the theory learned into practice and helps to cement the learning.

8.3.1.5 Demonstration of Skills

The demonstration of skills effectively represents a quality check on the candidate's ability to perform their job role. It involves an assessment of how well facility personnel have absorbed the information imparted to them during the training sessions. This is achieved by observing the candidate performing a given task as witnessed by an instructor. The instructor makes the assessment of their performance and provides constructive feedback.

Taking this a step forward, the training also should include an approval to verify that the candidate is capable of carrying out the given task. This may be in the form of a written test or skill demonstration, or both. There is usually a record of the assessment made, which is recorded and retained with the human resources (HR) group.

8.3.1.6 Emergency Response

The facility training program must also incorporate the process and actions to take in the event of an emergency situation at the facility, or *emergency response*. This may include the following components:
- shutdown of the facilities in a safe manner;
- first aid assistance for facility personnel;
- coordination and communication for affected parties at the facility, including emergency response groups;
- critical actions to take in the event of a facility toxic gas release;
- critical actions to take in the event of a facility fire.

8.3.1.7 Refresher and Training

As we have discussed, refresher training focuses on sustaining the high-performing team's knowledge and skills, basically to reconfirm competence.

Refresher training frequency intervals should be tailored to the training in question. It is also important to ensure that the training reflects company and legislative requirements.

8.3.2 Recordkeeping

Training records should be maintained within the facility HR group and should contain the following basic information: employee details; training material details; qualification tests and results, and trainer details.

Records should also distinguish between first-time training and refresher training and note all of the dates of the particular training conducted, along with respective upcoming refresher training needs.

A high-quality, effective training program depends on accurate identification of skills and knowledge gaps. The training program must be delivered across the facility organization in order to improve the overall understanding of facility integrity and best practices. The training program must be pitched at the right level to ensure that all staff are trained and developed to ensure the competence level is sufficient to match their corresponding facility job role requirements.

CHAPTER 9

Continuous Improvement

Contents

9.1 INTRODUCTION

We have now presented the 10 elements that make up the Facility Integrity Excellence Model (FIEM), which stem from the three core segments: Facility, Integrity Supporting Processes, and Personnel. These elements provide a basis for facility integrity excellence; however, there is an essential component that must also be present in order to achieve facility integrity excellence, and that is the requirement for continuous improvement (Figure 9.1).

"It is not possible to measure what you cannot control and you cannot control what you cannot measure!" [9.1]

Measuring performance takes a lot of effort, it can be expensive to set up and it has an ongoing cost to sustain. So why is it so important to measure the performance of facility integrity?

If we take a moment to analyze the key benefits associated with measuring performance, the argument is clear. Ultimately we measure performance because we want to know how good or bad we are at doing something. This in turn provides us with a basis to continuously improve our performance. We can then benchmark ourselves against industry norms and against our competition.

Additional benefits that may be achieved are as follows:

- *Provide a benchmark as to "where we are today"*: Performance measurement enables us to quantify our current position. How are we currently

Facility Integrity Management
http://dx.doi.org/10.1016/B978-0-12-801764-7.00009-7

Figure 9.1 *Continuous improvement.*

performing? This relates to all of the facility groups, including operations, maintenance and facility integrity and reliability (FI&R). Once we have a benchmark we are able to develop targeted and appropriate improvement programs. Benchmarking performance also helps us to secure the appropriate support from senior management for these potential improvement programs by providing a solid basis for justification.

- *We can make our achievements and accomplishments visible*: By publicizing our achievements through measured performance, we can inject a "morale booster" into the organization. This by nature also instills a sense of pride in our work and helps develop the right kind of culture.
- *We can compare the facility organization against common targets and objectives*: As the facility organization is working towards and being measured against the same goals, we can identify the top performers in the organization and help to support and develop their careers accordingly. In addition

we can identify employees in the organization who need additional support and provide it to them.

- *Highlight critical problems.* Data-driven scorecards can accurately identify where the critical problems are, which helps to focus the facility organization onto solving them.
- *Consistent performance measurement*: Once a robust and accurate performance measurement is up and running and facility data is collected and analyzed routinely, a powerful framework is created which enables a sound basis for decision making.

9.2 PERFORMANCE MEASUREMENT OVERVIEW

Continuous improvement is about learning from our past performance and making changes to improve our future performance. World-class facility integrity organizations are able to effectively set goals and targets and measure and track performance against those goals and targets. This is applicable to all of the facility functions, including maintenance, operations and FI&R, and all of the supporting function such as human resources (HR), management of change (MOC), finance, procurement, quality, health, safety and environment (HSE), etc.

This depends on our ability to:

1. select the appropriate measurement parameters that are relevant to each particular facility function and also the business objectives;
2. effectively collect robust and accurate data from the facility process and organization;
3. process the collected data to develop trends and highlight critical areas.

9.2.1 Integrity Performance Data Flow

We may refer to Figure 9.2, which presents a work flow to illustrate the process for continuous improvement in the facility. The continuous improvement workflow defines a set of key performance indicators (KPIs), which are an output of the overall business strategy for the facility and form part of the facility dashboard or *scorecard*.

Once the business strategy is set and objectives approved, the local facility strategy can be developed. This is usually prepared for the lifecycle of the facility, which is normally approximately 25 years. This 25-year facility strategy defines the overall targets and goals for the facility in terms of performance, such as production targets, maintenance budgets, reliability performance, facility availability, etc. These targets and goals are also

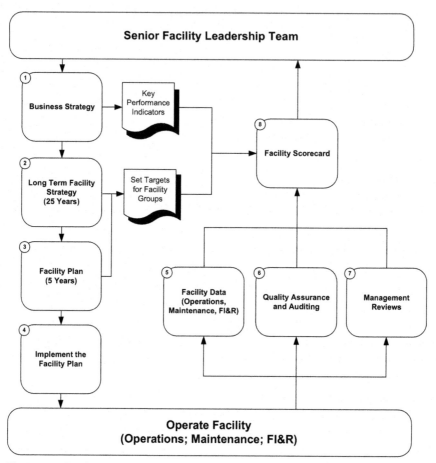

Figure 9.2 *Continuous improvement workflow.*

incorporated into the facility scorecard. The 25-year facility strategy is then translated into a 5-year operational plan which is implemented. Usually the plan is focused on 5 years because that is often the target interval between major facility shutdowns.

We then move into the operations stage of the facility lifecycle. We have touched on data collection previously in the course of this book. Data is collected from numerous sources within the main facility groups: operations; maintenance and FI&R. Typically data may include the following: technical performance data such as MTBF, reliability, and availability; workflow process performance data such as facility inspections performed against a plan, efficiency of maintenance work done, number of rework events, etc.; personnel data such as training conducted, etc.

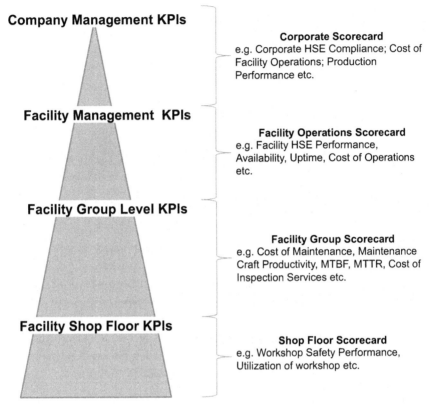

Company Management KPIs

Corporate Scorecard
e.g. Corporate HSE Compliance; Cost of
Facility Operations; Production
Performance etc.

Facility Management KPIs

Facility Operations Scorecard
e.g. Facility HSE Performance,
Availability, Uptime, Cost of Operations
etc.

Facility Group Level KPIs

Facility Group Scorecard
e.g. Cost of Maintenance, Maintenance
Craft Productivity, MTBF, MTTR, Cost of
Inspection Services etc.

Facility Shop Floor KPIs

Shop Floor Scorecard
e.g. Workshop Safety Performance,
Utilization of workshop etc.

Figure 9.3 *Hierarchy of KPI scorecards.*

Data is also collected through a number of the other FIEM elements including quality assurance and auditing and management reviews. These data are all collated and presented in the facility scorecard. The facility scorecard assesses the facility performance based on these data against the KPIs and targets and goals set by the business strategy and the long-term facility strategy.

The measurement of facility performance deals with a huge volume of data. The data must be assessed and filtered to ensure that the relevant details are presented to the relevant facility groups and hierarchy.

This is achieved by the development of a facility scorecard for each of the facility functions and at the different levels in the organization as illustrated in Figure 9.3.

It is common to develop KPI scorecards for the following groups as indicated in Figure 9.3:

• Corporate Team;
• Facility Management Team;

- Facility Group Teams;
- Shop Floor Teams.

The KPIs are usually recorded and assessed on a monthly basis by the individual facility functions, and reported up the line for management review, comment and action and down the line for the respective teams to respond to as required.

9.2.2 "What Gets Measured Gets Done"

"What gets measured gets done." This is an interesting phrase and suggests that if we take the time to measure our performance we will have confidence that work will be performed to meet set targets. Measuring the performance of organizations is standard practice for any modern organization. The process involves setting targets, measuring performance and acting on the results.

The process for developing useful KPIs is usually based on what has already happened in the past. These are called *lagging key performance indicators* and they are relatively straightforward to develop. It becomes more difficult when we try to identify and measure *leading key performance indicators*, which help to forecast future performance.

When going about developing a facility scorecard, it is important to have a blend of lagging and leading indicators so that we can learn from the past and can also have foresight of what may occur in the future.

9.2.3 Setting Key Performance Indicators

It is important to spend time up front to set up an effective and practical KPI scorecard. This is because the scorecard will require a lot of effort to maintain, update and report on periodically. KPIs that are defined in the scorecard will require certain information to be extracted from various sources, which may not prove to be practical or achievable. It is also in some cases not justifiable from a cost point of view to extract some information on a regular basis to use in the scorecard. We therefore need to consider what KPIs will add the most value and balance this with the effort and resources required to extract the information.

A typical process for setting KPIs is described as follows:

1. Decide and agree on the scope of the KPIs and scorecard. What are we trying to measure?
2. Select the organizational level at which the KPIs are aimed. The audience and interest in the KPIs differ considerably across the organization from the company corporate team to facility management to the various facility groups such as operations, maintenance, and FI&R.

3. Brainstorm, screen and select the key performance indicators that add value to that particular organizational level. Ensure that a blend of leading and lagging KPIs are defined.

4. Establish the mechanisms for extracting and collating the information and developing the KPIs. It is often effective to appoint a scorecard champion to extract the information from the various sources and effectively own the scorecard.

5. Establish the frequency of reporting.

6. Arrange the KPIs in a meaningful way and present them in a scorecard.

7. Obtain review and approval of the scorecard by the relevant representatives from the organization level selected.

8. Continually review the scope of the scorecard to ascertain whether the KPIs are still relevant and adding value.

The facility integrity scorecard should consider KPIs to address both "results" of the facility process performance-based KPIs, such as mean time between failures (MTBF), availability, uptime, etc. as well as a measure of how well the facility organization is performing facility-related activities or "internal" workflow-based KPIs. Examples may include compliance of planned versus actual maintenance work orders performed, rework activities for equipment overhauls, and so forth.

9.2.4 Facility Integrity KPIs

Facility integrity KPIs are developed by the numerous facility integrity groups and approved by the facility management team. KPIs should be tailored to meet the specific requirements of each individual facility in order to add value.

If we take a moment to review the key FIEM objectives, we can better understand and consider the KPIs that may be selected to measure the integrity of the facility. We discussed in Chapter 1 that the facility integrity management should:

• Maximize the availability of the facility;
• Provide integrity assurance for the facility;
• Provide a cost-effective facility integrity operation.

This means that the facility should be available as required, that the facility groups (maintenance, operation and FI&R) and supporting process are consistently functioning effectively and efficiently, all resources are optimized, with a focus on minimizing cost and having no integrity incidents.

We may now consider the specific Integrity KPI categories that may be employed in the development of a scorecard that can be used to effectively measure facility integrity performance.

9.2.4.1 Minimizing Cost KPIs

Minimizing cost is about applying the appropriate amount of effort to integrity-related activities to achieve optimal facility integrity performance. If we do not manage the cost of facility integrity management effectively and are not focused on cost, we may end up in a situation which presents an economically unviable operation. We are also not likely to achieve 100% availability in this unfocused way. On the other hand, if we spend no money on facility integrity management, there can be no integrity work performed and the facility operation will cease.

9.2.4.2 Execution of Integrity Activities KPIs

KPIs related to the execution of integrity activities address internal performance indicators and describe how well these activities have been performed. Internal workflow performance measurements in this sense may be based on:
- "Correctness" or accuracy of performing the execution activity – For example, has the work been performed correctly and in full in relation to the task plan?
- "Effectiveness" – How effective the process is at achieving its objective.
- Timely execution of work – Has the work been performed in line with the plan? For example, the number of planned inspection activities performed on time.
- Quality compliance – Were there any quality concerns identified during or after the integrity execution activity?
- Cost effectiveness – Is the activity done in a cost-effective way?

9.2.4.3 Integrity Performance KPIs

KPIs that focus on identification of the early onset of equipment and system failure (incipient failures) are helpful in addressing potential catastrophic failures ahead of time. Integrity performance KPIs may include indication of equipment malperformance: for example, number of critical pumps operating outside of the operating envelope. KPIs set up in this way may also give leading indication of pending equipment failure.

9.2.4.4 Integrity Incident KPIs

The tracking of integrity-related incidents are a major focus in any facility integrity scorecard. Integrity-related incidents are tracked through the incident reporting and tracking system, by the respective facility groups. The obvious target is to eliminate all incidents – "zero failure." Measuring integrity incidents provides indication of historic performance so it is a lagging indicator. It is a measure of how effective integrity management activities have been in the past. It does not give indication of future performance.

In order to emphasize the critical integrity-related incidents, the individual integrity incidents are referenced with their potential consequences. This immediately highlights which integrity incident-related KPIs to focus on.

9.2.4.5 Compliance KPIs

Compliance KPIs measure the compliance of performing integrity activities against a set standard: for example, carrying out on-time maintenance tasks, inspection activities, root cause analysis activities, operations rounds, etc. The measure also extends to an assessment of the adequacy and utilization of resources, materials, tools, timing and budget allocated, etc. against a preset standard.

Deviation in measured performance of compliance-type KPIs may be due to a number of reasons, including poor planning, mistakes made during the work order development, the ineffectiveness and inefficiency of the preset standards, etc.

One of the key factors to consider when setting compliance-type KPIs is that they provide no indication of how effective the standard set is. This factor should be addressed by the facility management team during the management review workflow process.

Compliance KPIs also do not provide any detail associated with the identification of the specific causes of poor compliance scores. It is important to cross-reference compliance-type KPIs with others in order to arrive at a more detailed picture of the overall performance. This can be achieved by careful design of the KPI scorecard.

9.3 INTEGRITY PERFORMANCE SCORECARD

The correct application of the FIEM will result in the integrity performance of the facility fully meeting the expectations of the corporate and facility management teams. The required performance measurement criteria, which is used as the basis for assessing the performance of the FIEM should be established and reviewed on an annual basis. This is because of the nature of the changing environment of numerous elements that make up facility integrity.

An integrity performance scorecard is developed to capture and arrange the specific performance KPIs for reporting to the various stakeholders. The scorecard is usually prepared and issued on a monthly basis.

An example of a Facility Integrity Excellence Model scorecard is shown in Figure 9.4. It is noted that integrity scorecards must be tailored to meet the specific needs of each facility and therefore Figure 9.4 is only indicative. The scorecard should also include a benchmarked target for each KPI, which is not shown in Figure 9.4. Scorecards similar to this

Facilities Integrity Excellence Scorecard	Target	Jan	Feb	Mar
Overview				
Total integrity incidents (No.)				
Availability (%)				
Availability of Critical Equipment (%)				
Uptime (%)				
Downtime(%)				
Maintenance Function				
Total work orders raised (No.)				
Total work orders completed (No.)				
Total work orders completed on time (No.)				
Total reactive work orders (No.)				
Total 'Priority 1' work orders (No.)				
Percentage of reactive work versus proactive work (%)				
Schedule compliance (percentage of completed / planned work orders) (%)				
Total back log work orders (No.)				
Maintenance Cost as a percentage of total Facility replacement value (%)				
Total ongoing maintenance cost ($)				
Cost of failures ($)				
FI&R Function				
Mean Time Between Failures (MTBF)				
Mean Time To Repair (MTTR)				
Total failures recorded (No.)				
Total 5 why investigations conducted (No.)				
Total failure investigations conducted (No.)				
Total failure investigations conducted / total failures recorded (%)				
Total rework events (No.)				
Total number of equipment items being monitored (No.)				
Operations				
Total operation unscheduled shutdowns (No.)				
Total Pre-Start Up Safety Reviews conducted (PSSR) (No.)				
Percentage of PSSRs / unscheduled shutdowns (%)				
Total cost of production loss ($)				
Supporting Processes				
Total MoC proposals raised (No.)				
Total MoCs approved (No.)				
Overdue modification forms (No.)				
Total quality inspections performed / quality inspections planned (%)				
Total overdue quality audits (No.)				
Total number of Facility personnel trained				

Figure 9.4 *Facility KPI dashboard.*

will be prepared for the different groups at the facility, as illustrated in Figure 9.3.

9.4 INTEGRITY PERFORMANCE SUSTAINABILITY

9.4.1 The Case for Management of Integrity

The need to assure integrity over the life cycle of a facility, from facility concept design to final abandonment, is one of the key factors to control the risks associated with deviations in integrity performance of the facility.

From facility concept design there is a need to define both the criteria and the risks associated with managing integrity. It is then pertinent to establish the controls to mitigate these risks. The influence of controls on the life cycle of a facility is shown in Figure 9.5, which illustrates the influence over the life cycle.

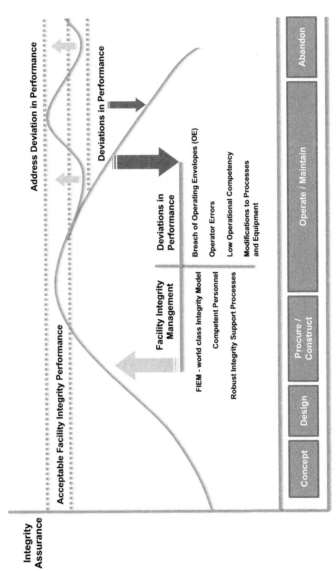

Figure 9.5 *Sustaining facility integrity performance.*

Integrity performance will decline with time due to gaps in critical management systems by events and circumstances such as:

- Breach of Operating Envelopes (OEs);
- Operator Errors;
- Low Operational Competency;
- Uncontrolled Modifications to Processes and Equipment.

Any relaxation of control over the operational life of a facility influences the robustness of systems, processes and resources deployed to manage integrity. The consequences can be severe, as we have seen in the case study in Chapter 2.

Integrity excellence controls are developed right at the start of the facility life cycle. Over time, confidence is amassed which ultimately leads to the establishment of a control scorecard to maintain integrity over the remaining period, as indicated by the "Acceptable Facility Integrity Performance" limits in Figure 9.5.

To restore the integrity performance to a satisfactory status, it is necessary to identify the gaps and close them. Integrity performance will again start to decline with time following the initial assessment, unless we institute a sustainability mechanism to enable us to assess and identify gaps on a continuous basis to sustain the integrity of our facility.

With robust integrity controls in place we can confidently sustain high integrity performance. The FIEM provides a toolbox to enable sustainability, including competent personnel and risk-centered culture (RCC), robust integrity support processes, and integrated facility groups all working towards a common goal.

CHAPTER 10

Implementation

Contents

10.1 INTRODUCTION

Change is a critical part of life within the oil and gas and petrochemical industries. If we refer back to Chapter 1, we recall that the approach to the management of integrity has been driven to change in line with the changing industrial landscape. For this reason we have developed an all-encompassing model to ensure facility integrity excellence, the Facility Integrity Excellence Model (FIEM). The implementation of the FIEM will impact the far reaches of the facility organization and each individual working at the facility.

Embarking on a journey of change that affects the entire facility organization is no small feat. In order for the change to be effective and to ensure the change is accepted by the organization, it must be well planned and executed carefully. The implementation of FIEM is no exception.

Facility Integrity Management
http://dx.doi.org/10.1016/B978-0-12-801764-7.00010-3

The implementation of a major change in the facility will influence the objectives of each facility group and the way each group goes about achieving those objectives. It will also influence how the facility organization shares information and communicates between the numerous facility groups.

The workflow process shown in Figure 10.1 provides a snapshot of some of the key elements that need to be considered during the implementation of FIEM into an organization. This change involves a wide range of functional groups, such as facility integrity and reliability (FI&R); maintenance operations; health, safety and environment (HSE); and quality; as well as processes and systems in the facility such as policies, procedures, standards, and information technology systems.

Effective implementation of the FIEM requires the nomination of champions who have a key role to play in the implementation process. These champions must be carefully selected and act as custodians of their respective facility groups: FI&R, maintenance, operations and the integrity supporting processes. Their role is to provide leadership to their respective facility group, promote FIEM, and drive the implementation of the FIEM through an appropriate change management framework in order to achieve the facility integrity objectives.

The facility senior management team also has a key role to play in the implementation of the FIEM. Senior management support is a fundamental success factor for any major change initiative. If senior facility management does not buy into the change and provide their support, the rest of the facility organization will follow suit. Senior facility management must "walk the talk." They must set the stage for the entire facility organization to follow, providing leadership from the top. They also must provide guidance for the facility group champions and the rest of the facility organization, in order to successfully deliver the change.

10.2 CHALLENGES IN CHANGE

With the implementation of the FIEM, as well as with any other major change, there will be a major shift from the existing long-standing facility practices. This in turn will present a number of key challenges within the facility organization and it is important to develop plans for managing the change. Some of the key challenges in change are presented here.

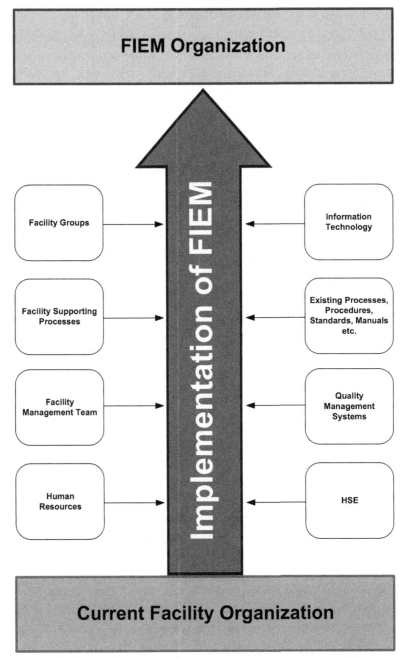

Figure 10.1 *The impact of organizational change.*

10.2.1 Ownership and Accountability

As with all processes, an implementation process needs to have an owner in order to ensure accountability for delivery. It is important to provide clear roles and responsibilities for each of the different team members associated with the implementation of the FIEM. The FIEM will likely create new roles as well as catalyze the development of existing roles. These new and updated roles should be clearly detailed in the individual job descriptions along with specific integrity-related key performance indicators (KPIs) to instill individual accountabilities to the FIEM.

The implementation of FIEM should be owned by the FI&R group with support from the other facility groups and functions. This is important so that input and updates to the relevant existing processes, procedures, standards and manuals are sufficiently covered. Usually a Responsible, Accountable, Consulted, Informed (RACI) chart, which is a popular management tool, is used to capture the ownership and accountabilities of the numerous stakeholders within the implementation plan.

10.2.1.1 Transition Management Teams

It is common practice to deploy a transition management team (TMT) when implementing a major change. The TMT is tasked with the smooth implementation of the change, the FIEM in this case. The TMT consists of a cross-section of stakeholders involved in the change effort. In the case of FIEM we have identified facility champions from the key facility groups, including FI&R, maintenance, operations and the integrity supporting processes.

The first step along the way for the TMT in the implementation process is to prepare a detailed implementation plan, which identifies the key objectives for the delivery of the change initiative. Often for a major change, it is pertinent to implement the change in phases.

The implementation plan may start with one of the facility processing units, for example. Once the change is implemented properly, the TMT will focus on another processing unit and so on. In some cases it may also be worthwhile to conduct a trial for a particular element of the FIEM to provide an opportunity to see how effective the change management effort is. In this way there is a chance to correct any issues within the FIEM implementation efforts prior to the full facility roll-out.

10.2.2 Communication

Effective communication during the implementation of a major change is a key challenge in change management. Poor communication will likely end up with a resistant-to-change organization; this is a common cause of failure of change initiatives. If the facility organization is not included or is unaware of the change, the implementation plan may not proceed smoothly.

Good communication is an essential ingredient to a successful change management program [10.1]. Given this, special attention should be applied to the onboarding plan for the implementation effort. A number of considerations that may be included within the onboarding plan are as follows:

- Introduce a routine (weekly or monthly) communication to the facility organization to keep them abreast of the change initiative progress. This may be in the form of a newsletter, email correspondence, website broadcast, etc.;
- Introduce a routine presentation to the facility organization to provide face-to-face communication and an opportunity for questions and answers;
- Introduce a change management website or web page(s) specifically for the change initiative. This media can include a host of information that may be consulted by the facility organization personnel at their convenience. The website may also include useful links for help guides to using new processes and systems, for example;
- The onboarding process may also include interviews to ascertain the opinions of the facility organization in the case of the FIEM. The FIEM implementation program may then be tailored to address any concerns and win the confidence of the facility organization;
- An onboarding plan should consider the celebration of successes along the way to achieving the ultimate goal of the implementation plan. This provides much-needed momentum to carry the change forward. Successes may be in the form of successful partial implementation of the initiative on facility processing units. Praising the facility team for their successes associated with the change initiative could be achieved by publishing articles in the company magazine, for example.

10.2.3 Resistance to Change

Early onboarding of the facility organization is crucial for an effective implementation process. It is often the case that the less effort we put into onboarding the organization, the more resistance they are likely to have towards the change. This is usually because, if the onboarding effort is weak, the organization has limited involvement and therefore does not feel part of the change effort. Organization personnel are more likely to feel that the change initiative is a force fit and therefore feel resistant towards it.

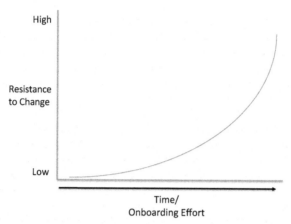

Figure 10.2 *Resistance to change.*

Figure 10.2 illustrates the key point that the longer we leave the onboarding process and the less effort we put into it, the more resistant the recipients, the facility organization, are likely to be to the change.

This relationship may vary depending on a number of factors, such as culture, organizational maturity, etc. and should therefore not be taken literally; however, Figure 10.2 provides an indication based on experience of the importance of a robust and timely change management onboarding plan.

10.2.4 Integration with Existing Company Processes

The implementation effort must also ensure that all of the elements of the FIEM are integrated with the various existing facility systems and processes where applicable. This may also include existing procedures, standards and manuals. The intention is to ensure that integrity management is engrained into the existing facility infrastructure from the bottom to the top.

Through the careful integration of the FIEM into the existing organizational systems and processes, the FIEM naturally becomes an integral part of the way the facility achieves the business objectives and addresses risk. It also eliminates the potential for inconsistency and duplication of standalone facility integrity systems. The implementation plan must carefully consider the management of knowledge and in particular the flow of facility integrity information (Figure 7.7).

10.2.5 Lack of Training

If a change initiative requires new training and none is provided, it is inevitable that there will be resistance to the change and furthermore a sound basis for the facility organization to resist change will have been created. Training is a crucial enabler to ensure the successful implementation of a new change within the facility organization. The training program must be tailored to meet the needs of the specific facility organization.

10.3 AN EFFECTIVE MODEL FOR THE IMPLEMENTATION OF CHANGE

Change is becoming a way of life and it is important that we embrace this reality and ensure we are equipped to deal with it. Planning for the implementation of change requires careful and detailed analysis of the current situation in order to develop an implementation plan that targets the key issues. This enables the change to be managed in a methodical way and therefore avoids reacting to a superficial assessment of the situation. Careful planning ahead of the implementation of a major change minimizes the resistance to change and often results in a successful change management program.

There are a number of change management theories that can be applied to the implementation of FIEM. One of the leading tried and tested change management theories is by John Kotter [10.2]. Kotter has developed a practical and effective eight-stage process for leading change.

A workflow process to illustrate Kotter's eight-stage process for leading change is shown in Figure 10.3.

The successful implementation of major change initiatives progresses through all eight stages. This is normally done in the sequence shown in Figure 10.3. However, in practical terms the implementation of change usually progresses through several stages at once or even misses a stage due to the pressures of delivering change. However, missing even a single stage in the process often produces difficulties.

The eight stages of Kotter's model are described in the following sections.

10.3.1 Establishing a Sense of Urgency

Establishing a sense of urgency is central to obtaining much-needed cooperation for the implementation of the FIEM. If the sense of urgency is low it is often difficult to develop and form a TMT with enough power and credibility to deliver the change initiative or to influence the key stakeholders to buy into and create the new change vision. It is important that the TMT is a highly competent and experienced delivery team. If there are high levels of incompetence, change initiatives usually fail, as few people

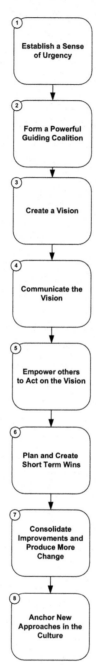

Figure 10.3 *Kotter's eight-stage process for leading change.*

are interested in being part of them. The sense of urgency must awaken the organization to believe and feel the need for change and therefore buy into the new change.

10.3.2 Forming a Powerful Guiding Coalition

A powerful guiding coalition is needed to drive the change initiative forward and maintain the momentum to ensure full implementation and effective on boarding of the facility organization. The guiding coalition must have the right team members, as we have discussed with the TMT in order to develop the appropriate level of trust within the organization. Building such a team to deliver change is a crucial part that must be done during the early stages of any major change initiative effort.

10.3.3 Creating a Vision

The implementation of major change initiatives can be set about in a number of different ways. All too often senior management attempts to implement change in their organizations by micromanagement or authoritarian decree.

Authoritarian decree rarely achieves its objective because it is unlikely to break through all the forces of resistance within the organization. Often the workforce tends to ignore the decree or pretends to cooperate while undermining senior management.

Micromanagement, on the other hand, avoids this problem by specifying in detail what employees should do and then closely monitoring them for compliance. This approach may break through some of the barriers to change; however; it is likely to be unacceptably slow.

Creating a vision for the future allows the facility workforce to appreciate the benefits of the change and buy into it on their own accord.

10.3.4 Communicating the Vision

The vision must be effectively communicated throughout the facility organization. It is also important that senior management are consistent in their approach to the vision in order for it to have credibility – as we have discussed, senior facility management must "walk the talk." They must set the stage for the entire facility organization to follow.

Caution must be taken to ensure the message of the vision is not lost or diluted because there is too much other communication occurring at the same time. The vision should be communicated in simple terms that are easy to understand and buy into by the entire facility organization. The message behind the vision should also be reiterated routinely to ensure it reaches the entire facility population. This can also be achieved using a number of different methods as we have discussed in section 10.2 (Challenges in Change).

10.3.5 Empowering Others to Act on the Vision

Empowerment is concerned with sharing information, rewards, and power with key facility employees so that they can take the initiative. It is all about removing obstacles for these key facility employees so that the entire facility organization can be engaged in the change initiative process. By empowering the facility organization, the change initiative can be embraced and further engrained into the organization.

10.3.6 Planning for and Creating Short-Term Wins

Before embarking on a major change initiative, it needs to be understood that the journey will take time. By planning and enjoying short-term wins along the way, the facility organization can see evidence that the change is bearing fruit and is worthwhile to stay the course. On the other hand, if we do not plan and create short-term wins, the nonbelievers may further convince themselves and others that the change is not a worthwhile commitment.

10.3.7 Consolidating Improvements and Producing Still More Change

We have established that we need to appreciate that the implementation of major change initiatives can take time. Along the way there may be many forces that can impede the process, including exhaustion on the part of leaders, the departure of key TMT members, etc. It is even more important under these circumstances that short-term wins are realized to keep momentum going. However, celebrating these wins can be destructive if the sense of urgency is lost, which will create opportunity for the forces of tradition to flood back into the organization. Therefore, it is essential to periodically consolidate the change improvements and continue to produce still more change in line with the implementation plan.

10.3.8 Anchoring New Approaches in the Culture

We have discussed risk-centered culture in detail in Chapter 3. Culture refers to the norms of behavior and shared values among a group of people. Norms of behavior are common ways of acting that are found in a group. They persist because group members tend to behave in ways that teach these practices to new members. Shared values tend to shape group behavior and

FIEM Implementation Plan											
No.	Equipment Description	Facility Equipment Number	Criticality Ranking	Equipment Maintenance and Operating Plan (EMOP)	Equipment Maintenance and Operating Card (EMOC)	EMOC Install	Pump Plinth Painted to signify Criticality	PMO / RCM Complete	FMEA Study	Spares Holding Optimised	Surveillance Round Detailed	...
1	Pump No.1	P121	High	X	X	X	X	X	X	X	X	X
2	Pump No.2	P122	High	X	X	X	X	X	X	X	X	X
3	Pump No.3	P123	High	X	X	X	X	X	X	X	X	X
4	Pump No.4	P124	Medium	X		X		X	x	X	X	X
5	Pump No.5	P125	Medium				X			X		
6	...	...	...									

Figure 10.4 *FIEM implementation plan.*

they often linger over time even when group membership changes. Usually shared values are more difficult to change than norms of behavior because they are less visible, but more deeply rooted in the culture. When the ways of doing something become clear during the change initiative and are found not to be harmonious with the existing culture, they will be exposed to reversion.

The change initiative is destined to fail if the associated new approaches are not anchored in the facility culture. The challenge with this step of the model is to identify any incompatibilities with the change initiative or new vision and the existing culture. Once identified, effective ways must be found to engrain the new vision into the old roots.

10.4 TRACKING THE CHANGE

It is important to track the implementation progress of the FIEM so that we can evaluate and understand how the change effort is progressing. This will in turn enable us to gauge our performance on an ongoing basis and to make adjustments to the program based on our implementation performance. Measuring the implementation effort must be based on the implementation plan which sets the objectives and timescales for implementation.

Tracking the implementation progress of each of the key elements of the FIEM against the implementation plan can be done through a simple KPI tracker as shown in Figure 10.4. Figure 10.4 serves as an illustration only and must be tailored to each specific facility requirement. A simple traffic light system indicating the progress of implementation of each element of the FIEM processes may provide a sufficient basis for this purpose.

The measurement of implementation performance can also be evaluated periodically through self-assessment and audit.

CHAPTER 11

Facility Integrity Strategy

Contents

11.1 INTRODUCTION

The term *strategy* tends to deal with the development of a plan to achieve the business objectives and in doing so, the allocation and optimisation of resources, whether personnel, financial or physical.

In the context of the FIEM, strategy deals with delivering integrity excellence through the optimization of facility resources and implementation of world-class processes and systems. Facility integrity strategy also recognizes that the individual component parts associated with the management of integrity depend on each other to be successful, and it aims to leverage their associated synergies.

11.2 THE HIERARCHY OF INTEGRITY MANAGEMENT

In order to put integrity strategy into perspective, a hierarchy of integrity management systems is shown in Figure 11.1. The leadership team must provide support and direction for the facility managers to develop and implement an integrity management system that meets and aligns with the overall business strategy and regulatory requirements.

The hierarchy starts with a vision for integrity. A vision is essentially a high level statement of what we would like to achieve in the future, which serves as a guide for deciding on the future course of action. The vision is set by the company leadership team.

Once the vision is set, the business objectives can then be defined, adding depth and direction. The business objectives also define specific targets and focus for the facility organization.

The FIEM sets the strategy through the various elements that integrate together to form the model. These elements provide workflow processes, methodologies and ultimately facility procedures. The FIEM policy is approved by the facility senior management team. The primary goals of the facility policy are to:

- Make sure that the risks associated with the failure of facility equipment and systems are identified and managed appropriately;
- Ensure consistent and effective reporting of near misses and integrity incidents;
- Ensure no harm to personnel and a safe environment to work in;
- Effectively manage all changes in the facility through the management of change (MOC) process;
- Ensure the concept of equipment criticality is effectively developed, implemented and administered and that all critical equipment are clearly identified and managed accordingly;
- Ensure that all facility equipment operate within its corresponding operating envelope (OE) and that any breaches are managed through MOC;
- Ensure that the roles and responsibilities associated with integrity management are clearly defined and communicated;
- Ensure that all facility equipment failures have thorough root cause analysis or 5 why investigations conducted and the lessons learned are communicated throughout the facility and acted on;
- Manage facility integrity performance through effective facility key performance indicator (KPI) scorecards;
- All FIEM processes are audited and reviewed regularly so that they can be continually improved.

Since FIEM approaches integrity from a holistic perspective, it touches all of the major aspects of integrity management. It involves all of the facility groups: facility integrity and reliability (FI&R), maintenance, operations, and all of the supporting processes facility groups.

11.2.1 Integrity Strategy Delivery Tools

The FIEM has been developed in such a way that it provides a toolbox of integrity delivery tools in order to ensure the strategy is implemented and sustained. Some of the key tools are noted as follows:

- Reliability Centered Maintenance (RCM);
- Risk Based Inspection (RBI);
- Instrument Protective Functions (IPF);
- Integrated Corrosion Management Systems (ICMS);
- Variance Tracking System;
- Management of Change Systems (MOC).

Figure 11.1 *The hierarchy of integrity management.*

Facility integrity procedures, processes and systems provide guidelines for the effective management of facility integrity of the equipment and systems. This is represented in the final stage of the hierarchy of integrity management in Figure 11.1. These procedures along with the equipment maintenance operation plans (EMOPs) serve as primary integrity documentation that details the working processes necessary to ensure compliance with the strategy and the policy.

11.3 FACILITY INTEGRITY STRATEGY WORKFLOW

Figure 11.2 provides an overview of a workflow for the development of the facility integrity strategy.

At the highest level in the organization the strategic direction of the business is set for integrity management and the associated strategic business objectives defined. This is achieved with directions from the company shareholders as well as alignment with the overall company vision.

The facility strategic direction can then be defined for the operating life of the facility. This is usually 25 years of operation. It is commonplace, however, for facility operating cycles to be extended well beyond the operational design life. In doing so, reliance on the FIEM and similar integrity management systems is crucial. When strategic decisions are made to prolong facility life, operating companies must be confident that the integrity of the facility is in line with these plans. Inputs into the long-term facility objectives may include industry best practices and the equipment criticality register.

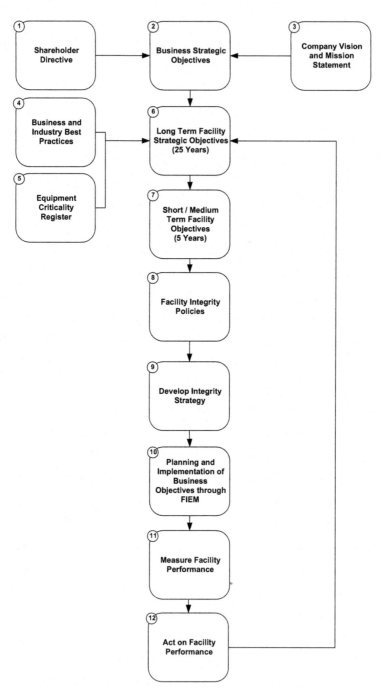

Figure 11.2 *Facility integrity strategy workflow.*

Once the 25-year operating plan and objectives are agreed upon, the facility management team is able to develop a detailed plan and objectives for approximately a 5-year time scale. This duration is usually selected for the facility because it is influenced by the time between major shutdowns. It is desirable to develop the strategy to prolong the time between major shutdowns beyond 5 years. The 5-year plan includes shutdown requirements for major overhauls as well as target metrics for facility integrity, such as availability, reliability, mean time between failures (MTBF), and reactive vs. proactive maintenance. The plan includes a detailed list of objectives to achieve those targets.

Once the objectives have been agreed, the facility group specific policies can be developed. Examples of some of the key policies are as follows:

- *Facility process safety policy.* Facility process safety deals with the identification of major hazards and develops measures to eliminate or mitigate the hazards. This policy should incorporate safe operating envelopes for facility equipment and corresponding control measures.

- *Facility maintenance policy.* The facility maintenance policy sets out how the maintenance organization is managed in order to minimize the risks of unscheduled downtime for facility equipment and systems. The maintenance policy also deals with ensuring that there is an appropriate balance between the various maintenance strategies, including condition directed, time directed and run to failure (RTF), as required. This is so that the maintenance resources and critical maintenance spares holdings may be optimized as much as possible and so that the facility availability performance is driven towards "best in class."

- *FI&R inspection policy.* The FI&R policy addresses both inspection and reliability requirements. In terms of inspection, the policy includes identification of the specific facility equipment for inspection and monitoring, the types of equipment and specific areas for inspection. It looks at how the equipment is to be inspected (on line or off line, invasive inspections or noninvasive techniques, etc.) and the frequency of inspections. The reliability policy addresses the identification and mitigation of potential equipment failures as well as the root cause analysis of failures in order to improve overall facility reliability. Management of facility equipment data (collecting, storage, updating, sharing, and so forth) is essential in order to address the issues of degraded equipment and the life extension assessment for continued service.

- *Competence policy.* The competence policy addresses the competence and skills of the facility organization. This relates to the requirements for skills, knowledge, experience and attitude of facility personnel, ensuring that each role in the integrity organisation is filled with the right personnel. It also focuses on how facility personnel are developed to ensure that the facility organization has the right competence level.

Following the development of the facility policy, attention is now turned to the development of the facility integrity strategy. The FIEM provides a robust framework in order to develop a strategy for the effective management of facility integrity, since it comprises all of the key integrity management elements. The strategy must also consider the facility integrity organization and its responsibilities in terms of competency and training, which we have detailed in Chapter 8. Once the strategy is developed, reviewed and approved by the facility management team, we are ready to implement it. The implementation process can be referred to in Chapter 10.

Finally, it is important to ensure that our performance is measured so that we can act on the results. Measuring our performance is set about through the detailed FIEM workflow processes and involves monitoring FI&R equipment trends, variance tracking, competence assessments, management reviews, and audits to name a few approaches.

The data collected is compiled through the facility scorecards, reviewed and acted upon, as described in Chapter 9 as part of the continuous improvement process.

Ensuring facility integrity is a cradle-to-grave activity. The management loop, which acts on facility performance ensures that the FIEM as a whole is working as intended and determines when appropriate changes and improvements are required to be made.

REFERENCES

[1.1] Public report of the fire and explosion at the ConocoPhillips Humber Refinery on 16 April 2001, Health and Safety Executive, UK, http://www.hse.gov.uk/comah/conocophillips.pdf

[1.2] Kyoto Protocol (2015). Website available at http://kyotoprotocol.com/default.aspx

[1.3] Macalister, T., Piper Alpha Disaster, *The Guardian*, Thursday 4 July 2013. Available at: http://www.theguardian.com/business/2013/jul/04/piper-alpha-disaster-167-oil-rig

[1.4] Nypro Chemical Plant at Flixborough in the UK, Health and Safety Executive, UK http://www.hse.gov.uk/comah/sragtech/caseflixboroug74.htm

[1.5] Bow Tie Model for Hazard and Effect Analysis, ICI (1970), http://en.wikipedia.org/wiki/Regulatory_Risk_Differentiation

[1.6] Kotter JP. Leading Change. Cambridge, MA: Harvard Business Review Press; 1996.

[2.1] Deming EW. Out of the Crisis. Cambridge, MA: The MIT Press; 2000.

[2.2] Public report of the fire and explosion at the ConocoPhillips Humber Refinery on 16 April 2001, Health and Safety Executive, UK, http://www.hse.gov.uk/comah/conocophillips.pdf

[2.3] Control of Major Accident Hazards, COMAH Regulations, Seveso II Directive in Great Britain Health and Safety Executive, available at http://www.hse.gov.uk/comah/

[2.4] Environment Agency (EA) UK (2015), website available at https://www.gov.uk/government/organisations/environment-agency

[3.1] British Standard, BS4778 – Risk Management Definition, British Standard, BS4778.

[3.2] Public report of the fire and explosion at the ConocoPhillips Humber Refinery on 16 April 2001, Health and Safety Executive, UK, http://www.hse.gov.uk/comah/conocophillips.pdf

[3.3] Kotter JP. *Leading Change*. Cambridge, MA: Harvard Business Review Press; 1996.

[4.1] Moubray J. P-F failure curve. Reliability-centered maintenance. 2nd ed. Industrial Press; 1997.

[4.2] Nowlan FS, Heap HF. Bathtub curve. Reliability-centered maintenance. Dolby Access Press; 1978.

[4.3] RCM Standard SAE JA1011, Society of Automotive Engineers, Standard JA1011, Evaluation Criteria for RCM Processes, available at http://www.sae.org/

[5.1] Arunraj, N.S. and Maiti, J. Risk-based maintenance: Techniques and applications, (Evolution of Maintenance), Journal of Hazardous Materials, Volume 142, Issue 3, 11 April 2007, pages 653-661.

[5.2] Nowlan, F.S. and Heap, H.F. DOD report number A066-579, Reliability-Centered Maintenance. United States Department of Defense. December 29th, 1978.

[5.3] Smith AM, Hinchcliffe GR. Preventive Maintenance. RCM—Gateway to World Class Maintenance. Elsevier; 2003.

[5.4] RCM Standard SAE JA1011, Society of Automotive Engineers, Standard JA1011, Evaluation Criteria for RCM Processes, available at http://www.sae.org/

[5.5] United States Department of Defense (9 November 1949). MIL-P-1629 – Procedures for performing a failure mode effect and critical analysis. Department of Defense (US).

[5.6] Standard J1739, Potential Failure Mode and Effects Analysis in Design (Design FMEA) SAE, Society of Automotive Engineers, SAE. http://www.sae.org/

[5.7] International Standard on Fault Mode and Effects Analysis, International Electrotechnical Commission (IEC), Standard IEC 60812, Fault Mode and Effects Analysis.

[5.8] Deming, W. Edwards (2000). The New Economics for Industry, Government, Education (2nd ed.). MIT Press. ISBN 0-262-54116-5. OCLC 44162616.

Facility Integrity Management
http://dx.doi.org/10.1016/B978-0-12-801764-7.00012-7

[7.1] ISO (International Organization for Standardization), ISO Central Secretariat, available at http://www.iso.org

[7.2] ISO (International Organization for Standardization), 19011:2011, ISO 19011:2011, Guidelines for auditing management systems. Available at http://www.iso.org

[8.1] Wagner, Robert W. Edgar Dale: Professional. Theory into Practice. Vol. 9, No. 2, Edgar Dale (Apr., 1970), pp. 89-95. Taylor & Francis, Ltd. http://www.jstor.org/pss/1475566

[9.1] Drucker Institute, 2015, "Why Drucker Now?", available at http://www.druckerinstitute.com/peter-druckers-life-and-legacy/

[10.1] Kotter JP, Cohen D. The Heart of Change: Real-Life Stories of How People Change Their Organisations. Harvard Business Review Press; 2002.

[10.2] John P. Kotter (1996). Leading Change, Harvard Business Review Press, Jan 1, 1996.

SUBJECT INDEX